Cellular Ageing

Monographs in Developmental Biology

Founded 1969 by *A. Wolsky*, New York, N.Y.

Vol. 17

Series Editor
H.W. Sauer, College Station, Tex.

S. Karger · Basel · München · Paris · London · New York · Tokyo · Sydney

Cellular Ageing

Editor
H.W. Sauer, College Station, Tex.

79 figures and 28 tables, 1984

QH608
C44
1984

S. Karger · Basel · München · Paris · London · New York · Tokyo · Sydney

The publication of this volume was made possible through a generous contribution by the Heinz Karger Memorial Foundation.

National Library of Medicine, Cataloging in Publication
　　Cellular ageing/Editor, H.W. Sauer. – Basel: Karger, 1984. –
　　(Monographs in developmental biology; v. 17)
　　Includes index.
　　1. Aging 2. Cell Survival I. Sauer, Helmut W. II. Series
　　W1 M0567L v. 17 [WT 104 C393]
　　ISBN 3-8055-3860-X

Drug Dosage
　　The authors and the publisher have exerted every effort to ensure that drug selection and dosage set forth in this text are in accord with current recommendations and practice at the time of publication. However, in view of ongoing research, changes in government regulations, and the constant flow of information relating to drug therapy and drug reactions, the reader is urged to check the package insert for each drug for any change in indications and dosage and for added warnings and precautions. This is particularly important when the recommended agent is a new and/or infrequently employed drug.

All rights reserved.
　　No part of this publication may be translated into other languages, reproduced or utilized in any form or by any means, electronic or mechanical, including photocopying, recording, microcopying, or by any information storage and retrieval system, without permission in writing from the publisher.

©　　Copyright 1984 by S. Karger AG, P.O. Box, CH-4009 Basel (Switzerland)
　　Printed in Switzerland by Thür AG Offsetdruck, Pratteln
　　ISBN 3-8055-3860-X

Contents

Foreword VII
Twenty Years Heinz Karger Memorial Foundation IX

I. Cellular Ageing: An Overview

Aufderheide, K.J. (College Station, Tex.): Cellular Ageing: An Overview 2
Kirkwood, T.B.L. (London): Towards a Unified Theory of Cellular Ageing 9
Harrison, D.E. (Bar Harbor, Me.): Do Hemopoietic Stem Cells Age? 21
Bowen, D.M. (London): Cellular Ageing: Selective Vulnerability of Cholinergic Neurones in Human Brain 42
Holliday, R. (London): The Unsolved Problem of Cellular Ageing 60

II. In vitro vs. in vivo Ageing

Flood, M.T.; Haley, J.E.; Gouras, P. (New York, N.Y.): Cellular Ageing of Human Retinal Epithelium in vivo and in vitro 80
Rink, H. (Bonn): Growth Potential, Repair Capacity and Protein Synthesis in Lens Epithelial Cells during Aging in vitro 94

III. Immune Senescence

Weksler, M.E.; Siskind, G.W. (New York, N.Y.): The Cellular Basis of Immune Senescence 110
Szewczuk, M.R.; Wade, A.W. (Kingston, Ont.): Cellular Aging, Idiotype Repertoire Changes and Mucosal-Associated Lymphoid System 122
Steinmann, G.G.; Müller-Hermelink, H.-K. (Kiel): Lymphocyte Differentiation and Its Microenvironment in the Human Thymus during Aging 142

IV. Gene Expression and Nuclear Ageing

Sarkander, H.-I. (Berlin): Molecular Mechanisms Decisive for Neuronal Ageing: A New Theory on Senescent Cellular Deterioration 158
Taş, S. (Gebze): Cellular Aging, Neoplastic Transformation, Meiotic Rejuvenation, and the Structure of Chromatin Complex 178
Smith, J.R. (Lake Placid, N.Y.): A Hypothesis for in vitro Cellular Senescence Based on the Population Dynamics of Human Diploid Fibroblasts and Somatic Cell Hybrids 193

V. Cytoplasmic Ageing

Spoerri, P.E. (Göttingen): Mitochondrial Alterations in Ageing Mouse Neuroblastoma Cells in Culture 210
Holliday, R.; Rattan, S.I.S. (London): Evidence that Paromomycin Induces Premature Ageing in Human Fibroblasts 221
Gazzola, G.C.; Bussolati, O.; Longo, N.; Dall'Asta, V.; Franchi-Gazzola, R.; Guidotti, G.G. (Parma): Effect of in vitro Ageing on the Transport of Neutral Amino Acids in Human Fibroblasts 234
Kay, M.M.B. (Temple, Tex.): Band 3, the Predominant Transmembrane Polypeptide, Undergoes Proteolytic Degradation as Cells Age 245
Cummings, D.J. (Denver, Colo.): Mitochondrial DNA in *Podospora anserina*. A Molecular Approach to Cellular Senescence 254

Subject Index 267

Foreword

On one level, ageing research can be viewed as one of those unusual cases where clinical success – rather than problems – has created the demand for basic research. Medical advances that have so successfully prolonged the human life span now leave us with a population suffering from ills distinctly related to longevity. While ageing has always been associated with failing health, it is the unprecedented size of this elderly population, a tribute to medical success, that now compels research in basic cellular and developmental biology to understand the mechanisms by which ageing alters health.

When the Heinz Karger Memorial Foundation decided to focus its 20th annual competition on 'Cellular Ageing', we expected a good response. Viewed at the cellular level, ageing research is again something of an unusual case. Perhaps due to the challenge inherent in its complexity, research in this field has attracted talented investigators using experimental and conceptual models drawn from some of the most promising areas of biomedical research.

We did not, however, expect to receive so many outstanding manuscripts. Nor did we imagine that these unsolicited papers would fall into natural groupings offering a fairly representative overview of the state-of-the-art in this field.

On June 30, 1983, the Foundation council decided to award the prize, in equal parts, to *T.B.L. Kirkwood* of London and *James R. Smith* of Lake Placid. The task of selecting the best among so much general excellence also proved more difficult than expected, and two weeks later, we decided to publish these prizewinning papers together with the 'runners-up'. This book is the result; the 'Monographs in Developmental Biology' series was selected as an appropriate context for its publication.

The publisher and other members of the council wish to thank *Helmut Sauer* for assuming editorial responsibilities and shaping these manuscripts into a unified form. *Karl J. Aufderheide*, whose contribution opens the book, accepted the elephant-sized task of placing this collection into the context of problems that now confront ageing research. Finally, we want to thank the many authors who updated or otherwise revised their manuscripts prior to publication.

We hope with this book, as well as with the award to Drs. *Kirkwood* and *Smith*, to have furthered the Foundation's goal of supporting excellence in basic research.

Dr. h.c. *Thomas Karger*

Twenty Years Heinz Karger Memorial Foundation

Dr. Dr. h.c. *Heinz Karger*
Director of S. Karger from 1935
until his death in 1959

The Heinz Karger Memorial Foundation was established in 1963 with the goal of honoring outstanding scientific research through financial prizes awarded on the basis of annual or biennial competitions. Throughout these years, the selection of winning papers – ranging from transport mechanisms in human leukocytes (1963) to catecholamines in the mechanism of anxiety (1973), and often anticipating broad acceptance of novel research strategies – has consistently been in keeping with *Heinz Karger's* belief that the best research is performed when freed from the need to produce immediately practical results.

In 1983, the Foundation marked its 20th anniversary with several departures from tradition. The topic, 'Cellular Ageing', was selected as representing an area fueled, and largely funded, by concern over the problems posed by an ever-larger geriatric population. Given the intensity of recent research in this area, the Foundation also relaxed its rule stipulating original findings only and allowed authors to submit review papers interpreting their personal contributions to the field. Finally, the financial prize was increased to more than double the amount of any previous year.

The 1983 competition also marked the first time that the Foundation recommended publication of selected manuscripts in book form. Though authors strictly adhered to competition rules when preparing manuscripts – a fact which expedited technical production – the topic of 'Cellular Ageing' was subject to individual interpretation on at least one small point: though everyone shares the experience of ageing, not everyone agrees on how to spell it.

Steven Karger

I. Cellular Ageing: An Overview

Cellular Aging: An Overview

Karl J. Aufderheide[1]

Department of Biology, Texas A&M University, College Station, Tex., USA

The study of the diverse collection of phenomena known intuitively as 'aging' is reminiscent of the old parable of the blind men and the elephant. Each blind man, having a limited ability to investigate only a portion of the elephant, comes away with a very different impression as to the nature of the elephant. Each impression is accurate within its context, but alone is insufficient to describe the elephant as a whole. So it seems to those studying aging. This metaphor is drawn, not in a demeaning spirit, but as a way of describing a fundamental difficulty of aging research: the lack of a paradigm (or a large-scale view of the elephant) providing a context within which specific questions may be asked, and specific answers may be understood. Scientists may await the coming of a Watson-and-Crick, or a Darwin, of aging who can provide such a paradigm. Or at least we hope that our intuitions are correct and that a paradigm of aging is possible. If it is not, then we may need to reconsider our definition of aging and whether the term has any scientific meaning (i.e., is there only one elephant, or could there be a number of more or less elephant-like creatures?).

The study of cellular aging, which may seem to be but a portion of the elephant, turns out to be another elephant in its own right, and no less difficult to summarize or describe. Rather than attempt such a description, it seems worthwhile for this collection of papers to ask a series of questions significant to cellular aging, and to discuss how the various contributors address those questions. Many of those questions have been eloquently discussed by *Kirkwood, Holliday* and others in this collection, so this article will, in many cases, only restate the ideas of the others.

---

[1] The author is supported by a grant from the National Institutes of Health – Institute of Aging, No. AG02657.

What Is the Significance of Cellular Aging to Organismal Aging?

A central concept of cellular aging is that many of the aspects of aging seen at the organismic level [7] are the results of aging of at least some of the cell populations of the organism [2]. This assertion can be tested by studying various cellular populations in vivo and their effects upon the organism as aging progresses. The difficulty in this kind of study is establishing a causal link between a particular aspect of aging and the aging of a particular cell population. The works of *Weksler and Siskind, Szewczuk and Wade*, and *Steinmann and Muller-Hermelink* upon the immune system and aging show a definite correlation between changes in the populations of immune system cells and the organism's loss of immune competency with age. *Bowen* has documented a correlation between a drop in the population of cholinergic neurons in the brains of humans and rats, and certain kinds of age-related senility. More studies of this type are needed, although it is usually not so easy to trace an aspect of an organism's physiology to a particular cell type. Especially needed are studies which can show a causal linkage between the loss or senescence of particular cell types and the appearance of certain aspects of organismic senescence. This last is required for a convincing demonstration that cellular aging is a major component or organismic aging.

Do Cell Populations Age in vivo as well as in vitro?

Early studies of cellular senescence used cells in tissue culture [1, 3]. These in vitro studies were susceptible to the criticism that the culture technique or the medium introduced artifacts not seen in vivo. It therefore became essential to document that cell populations were showing aging in the organism in ways similar to the aging seen in culture. In this collection of papers, the works of *Rink* and *Flood* et al. address the issues of in vitro versus in vivo aging. These works indicate that, in at least a qualitative sense, similar processes are seen in either case. Such a correlation is heartening, but the issue deserves broader study, especially for specific types of cells in specific portions of an organism. For instance, *Harrison* reports that stem cells may have longer proliferative life-spans than ever expected, and might be immortal. Another difficulty arises when one realizes that the proliferative span of many cell populations is longer than the life span of the organism from which the cells are derived [1, 2]. If the cells live longer than the organism, how can cellular aging contribute to organismal aging? Of course, we do not know in detail the conditions of various cell types in various portions of the organism, so it could be that a critical organ may develop a localized deficiency of a cell type, even though the whole body population of that cell type may be adquate. Experimental resolution of this apparent contradiction between in vitro culture proliferative span and the organismal life span is also needed to improve the correlation between organismal aging and cellular aging mentioned above.

Do Highly Differentiated Cells Show Aging?

Much of the work on cell aging has dealt with rapidly proliferating cell populations, which may tend to show fewer differentiated characteristics than tissues in vivo. The question may then be asked as to whether highly differentiated cells show aging phenomena, as well as the less differentiated cell types, such as fibroblast-like cells in culture. Or are aging effects seen only in rapidly growing cell populations?

In this collection of papers, *Sarkander* and *Bowen* both have investigated aging in neurons, which surely represent an extreme in differentiation, while *Rink* and *Flood* et al. investigated differentiated epithelial-type cells still capable of proliferation. In all 4 papers, aging effects are definitely seen in differentiated cells. The effects are particularly striking in neurons: *Sarkander* demonstrates a series of molecular changes in the neurons of the cortices of rats as they age. To answer the criticism that most of the brain consists of glial cells and not neurons, and therefore changes seen are largely glial cell changes, *Bowen* documents that specific loss of cholinergic neuron types as aging progresses. Thus, aging effects are not restricted to proliferating, undifferentiated cell populations, and the senescence of 'terminally differentiated' cell types also may have a significant influence upon aging in the organism.

What Are the Mechanisms of Cellular Aging?

Space does not allow a comprehensive review of this topic, and it is somewhat redundant to repeat the discussions of other contributors, especially the works of *Kirkwood* and *Holliday*. However, a discussion from a different perspective may be appropriate.

Theories of cellular aging can be classified into 2 groups, depending upon the mechanism postulated: theories where aging is the consequence of stochastic events and theories where aging is the result of programmed series of events. It is somewhat surprising that few of the contributors to this collection seem to display much enthusiasm for programmed aging theories. Rather, discussions in general tend to consider cellular senescence to be the accumulation of errors of cellular function, presumably generated by stochastic events. The relative merits of the 2 types of theories have not been resolved, though. The problem of cellular transformation, and the consequent escape from cellular mortality, becomes acute in this context. Since it is proposed that mortal cell lines lose their proliferative capacity and become senescent because of stochastic events, any theory of cell aging must be capable of explaining how a transformed cell line avoids these stochastic events and maintains its immortality. Many cancers are associated with chromosomal defects [13], which suggests that oncogenesis is itself the consequence of stochastic events. Furthermore, specific oncogenes are associated with transformation (and thus presumably with escape from mortality), suggesting that there are only a few specific paths to immortality [5]. *Smith* shows that the immortal state is recessive to the mortal state; this has been interpreted in favor of a programmed aging process, with stochastic events allowing escape from the program [6]. But, if immortality depends upon the expression of a few oncogenes, one might expect that transformed cells would be susceptible to damage of those oncogenes. This damage might be expected to cause reversion to the mortal state. Is there any evidence for this kind of reversion, or for an enhanced ability of transformed cells to protect themselves from stochastic events? *Kirkwood* suggests that immortal cell lines might have better repair mechanisms than mortal cell lines – this hypothesis should be tested, along with other comparisons of immortal versus mortal cell states. The resolution of aging mechanisms is thus far from complete.

Another approach to cellular aging mechanisms is to assess the relative contributions of the nucleus and the cytoplasm to the development of senescence in a cell population. Among the contributors, *Tas, Sarkander* and *Smith* find evidence implicating the nucleus as a source of aging. *Tas* and *Sarkander* find significant age-related changes in the chromatin which may hamper proper gene expression. *Smith* finds evidence suggesting genetic damage, or at least significant changes in gene expression. The accumulation of such damage would obviously lead to a condition where the nuclei could no longer provide adequate transcript for the life processes in a cell. Other contributors: *Spoerri, Holliday and Rattan, Gazzola* et al., *Kay* and *Cummings* find evidence for a cytoplasmic contribution to aging. There are various specific mechanisms proposed for cytoplasmic aging, but many depend upon the 'error catastrophe' model described by *Holliday*, in which errors are propagated through various compartments of the cell until one or more of the functions of life becomes impossible. As with many other aspects of aging research, the nucleus and the cytoplasm are so closely associated that it becomes difficult to separate the relative contributions of each to a particular event or process. Early cell fusion experiments [12], seeking to test various combinations of old and young nuclei with old and young cytoplasms, concluded that both the nucleus and the cytoplasm may contribute to the observed phenomena of cellular aging. More recent work, summarized by *Smith*, continues to support this idea, although nuclear factors may be of paramount importance. Also, the observations of *Cummings*, showing the transfer of a senescence-associated sequence of DNA from mitochondria to the nuclei of aging fungal cells, are especially interesting, since they indicate the possibility of nucleocytoplasmic interactions in aging.

In general, however, we are far from an agreed-upon mechanism for cell aging. There may be no one mechanism for cellular aging, but rather several different, perhaps independent, perhaps interdependent, processes all contributing to the observed cellular aging.

Other Issues

One approach missing from most of the papers in the collection (except that of *Cummings*) is the use of single-celled organisms as models for aging [8, 11]. There are many advantages to the use of protistan systems, but 4 are especially important: (1) the ability to use genetic and quasi-genetic manipulations and analyses [10] to dissect cellular aging – these are not easily done using tissue cultures; (2) the fact that protistan systems, with single cells in culture, are essentially in vivo, so that the potential problem of artifacts arising from cells being away from their normal conditions are essentially minimized; (3) since many protists are large cells, surgical manipulations, such as the transfer of nuclei or cytoplasmic volumes are possible [4], and (4) the large size of many species allows for individual cell selections for lineage studies [9] not easily done with tissue culture cells in mass. The author, a protozoologist, may have a particular bias, but some of the questions and issues discussed above are open to attack using one of the protistan species which do indeed show aging. Especially powerful is the ability to use the genetic systems of these creatures to assist in a better understanding of cellular aging [8]. If mutants conferring an especially short or long clonal life span can be found, then comparison of those mutants to wild-type cells could prove profitable to our understanding of some of the possible mechanisms of aging.

To return to the metaphor presented at the beginning of this article, it could be argued that, hearing the descriptions of the blind men, no such thing as an elephant exists. It is true that aging is an extremely difficult thing to define in a rigorous scientific sense, but it is also true that we intuitively seem to know what aging is. At least most of us can distinguish an aged human from a young one. But it is also likely that there is no single process of aging, but that rather the term is a convenient noun to describe a phenomenologically diverse set of processes producing more or less similar results. Such a situation would not invalidate 'aging research' any more than any other field of science which has expanded into more and more diverse disciplines as research continues. The various aspects of aging encompass most of what we consider to be biology, so research on aging will, in its sum, have a similar broad scope. It may be expected that at least some of the aging research will eventually benefit mankind, but which part this will be is uncertain. That is the risk and the exciting joy of research.

References

1 Hayflick, L.: The limited in vitro lifetime of human diploid cell strains. Expl Cell Res. *37:* 614–636 (1965).
2 Hayflick, L.: Cell biology of aging. Bioscience *25:* 629–637 (1975).
3 Hayflick, L.; Moorehead, P.S.: The serial cultivation of human diploid cell strains. Expl Cell Res. *25:* 585–621 (1961).
4 Knowles, J.K.C.: An improved microinjection technique in *Paramecium.* Expl Cell Res. *88:* 79–87 (1974).
5 Land, H.; Parada, L.F.; Weinberg, R.A.: Cellular oncogenes and multistep carcinogenesis. Science *222:* 771–778 (1983).
6 Pereira-Smith, O.M.; Smith, J.R.: Evidence for the recessive nature of cellular immortality. Science *221:* 964–966 (1983).
7 Reff, M.E.; Schneider, E.L.: Biological markers of aging; Conf. on Nonlethal Biol. Markers of Physiol. Aging, 1981 (National Institutes of Health, Bethesda 1982).
8 Smith-Sonneborn, J.: Genetics and aging in Protozoa. Int. Rev. Cytol. *73:* 319–354 (1981).
9 Sonneborn, T.M.: Methods in *Paramecium* research. Meth. Cell Physiol. *4:* 241–339 (1970).
10 Sonneborn, T.M.: *Paramecium aurelia;* in King, Handbook of genetics, vol. 2: Plants, plant viruses and protists, pp. 469–594 (Plenum Press, New York 1975).

11 Sonneborn, T.M.: The origin, evolution, nature, and causes of aging; in Finch, Moment, The biology of aging, pp. 361–374 (Plenum Press, New York 1978).
12 Wright, W.E.; Hayflick, L.: Nuclear control of cellular aging demonstrated by hybridization of anucleate and whole cultured normal human fibroblasts. Expl Cell Res. *96:* 113–121 (1975).
13 Yunis, J.J.: The chromosomal basis of human neoplasia. Science *221:* 227–236 (1983).

K.J. Aufderheide, Department of Biology, Texas A&M University, College Station, TX 77843 (USA)

Towards a Unified Theory of Cellular Ageing

T.B.L. Kirkwood

National Institute for Medical Research, Mill Hill, London, England

Introduction

The science of cellular ageing has, from its outset in the late 19th century, presented a wealth of conceptual problems. Early research was in fact largely confined to theorising and it was not until *Hayflick and Moorhead* [4] demonstrated that normal (i.e. diploid) human fibroblasts have finite division potential in vitro that model systems for experimental studies involving multicellular species became available. During the intervening years, the inspired ideas of *Weismann* [34], who was the first to suggest a cellular basis of ageing, were ignored or, for various reasons, discounted [for a review, see 12]. Now, as experimental data accumulate at an increasing rate, there is urgent need to establish a fresh conceptual framework into which new discoveries and ideas can be added.

Interest in cellular ageing stems principally from its relevance as a possible basis for ageing of the organism as a whole. In consequence, a coherent theory of cell ageing needs three components. Firstly, the nature and role of the ageing process at the organismic level must be explained. Secondly, a link between ageing at the organismic level and ageing at the cellular level must be established. Thirdly, the mechanism of cellular ageing must be defined. In addition, there is a need to explain why heteroploid transformed cells, unlike diploid cells, grow indefinitely in culture. This question has obvious relevance to cancer research and characterisation of the difference between 'mortal' and 'immortal' cultures is also likely to give considerable insight into the basic mechanisms of cellular ageing.

In this review, each of these basic questions is considered in turn, with special reference to the author's own recent research. A structural framework is proposed to accommodate current and future concepts on cellular ageing and it is suggested that efficiency of research effort may thereby be increased.

The Role of Ageing in Biology

Definition of Ageing

Ageing is, for the most part, defined in terms of its effect on survivorship. Species in which mortality increases at an accelerating pace during adulthood are generally said to age, whereas species in which mortality remains constant, or even declines, are held not to age. Apart from the practical difficulties of obtaining survivorship data adequate to establish presence or absence of ageing [11], this definition can usefully be applied to most higher animals.

The limitation to any single definition of ageing, such as that given above, is the diversity of organisms' life histories. Two classifications of life histories are of major importance here. The first is between species which have a clear distinction between germ cells and somatic tissue and those which do not. The second is between species which reproduce only once in their lifetime (semelparous) and those which reproduce repeatedly (iteroparous). The concept of 'ageing' is most clearly defined in iteroparous species which have a distinct soma, separate from the germ line, and may need to be considerably qualified when applied to species with other kinds of life history [12]. It is mistaken, for example, to regard the post-reproductive death of semelparous species, which usually occurs in highly determinate fashion, as being comparable with the more protracted senescence of iteroparous species.

An Evolutionary Perspective

Since the time of *Weismann* [34], biologists have tried to fit ageing into an evolutionary picture of life. The puzzle is to explain why the life span of higher organisms should be limited, when there seems no reason, in principle, why it could not be extended, perhaps indefinitely, and when most vital processes serve to prolong it [35]. It is perhaps natural, therefore, to suppose that ageing serves a useful purpose and that, in some subtle way, it is an evolutionary adaptation.

The view that ageing is adaptive is widely held and is implicit in much current gerontological research. In particular, it underlies the popular idea that ageing is a programmed process under its own strict genetic control. This suggests, in turn, that all that is required to understand, and possibly modify, the ageing process is to unravel the details of the programme. However, if it is to withstand close scrutiny, an adaptive theory of ageing must suggest a reason why, other things being equal, an organism which ages is fitter in a neo-Darwinian sense than one which does not. It is on this crucial test that adaptive theories have failed.

There are two complementary reasons why adaptive explanations of ageing are generally unsound. The first is simply that mortality in wild populations is usually so great that individuals rarely survive to senility [23]. Thus, there is neither need for any extra mechanism to terminate the life span, nor obvious opportunity for it to have evolved. The second reason, which counters the suggestion that ageing improves a species' capacity to adapt to changes in its environment by accelerating the turnover of generations [38], arises from considering the level at which natural selection operates. For ageing to have arisen adaptively as suggested would have required that selection for advantage to the *species* was more effective than selection among *individuals* for the reproductive advantages of a longer life. Superiority of the first kind of selection over the second occurs only under very special circumstances [22] and it is easily shown that an adaptive mechanism for ageing cannot normally be stably maintained [10]. This conclusion has sweeping consequences for all levels of gerontological research, which have yet to be fully recognised.

If ageing is not adaptive, then it has to be explained either by the failure of natural selection to prevent it, or as a by-product of other adaptive traits. In iteroparous species it is indeed the case that the force of natural selection diminishes with age [23, 35], which inevitably results in only loose control over the later portions of the life span. Of greater significance, however, is the fact that the life history will be organised to maximise reproductive success. A particular aspect of life history organisation is the trade-off between somatic maintenance and reproduction. The larger the fraction of resources invested in somatic maintenance, the smaller the investment in reproduction (and conversely), irrespective of whether resources are abundant or scarce. At either extreme, the organism pays a penalty. If somatic maintenance is too little, death will result very soon. If somatic maintenance is too much, reproduction is severely retarded. By mathematical modelling of the life history, it can be shown that the strategy of maximum fitness is one which balances somatic maintenance against reproductive effort in such a way that the investment of resources in maintenance and repair of the soma is *always less than what is required for indefinite somatic survival* [10; and Kirkwood, in preparation]. Thus, it is predicted that in iteroparous species the soma will always age through an accumulation of unrepaired somatic damage. This is termed the *disposable soma* theory of ageing [8, 10, 14].

It is instructive to consider closely how the disposable soma theory is related to concepts of adaptiveness and programming. First of all, the theory is truly a non-adaptive theory of ageing, even though ageing results from positive selection. The essential point is that ageing itself is regarded clearly as a disadvantage, which is balanced against the competing advantage of increased output of progeny. The positive aspects of selection deal with optimising the total outcome of the resource allocation strategy, of which senescence is a negative component. As for programming, the disposable soma theory predicts that the nature and average rate of senescence *are* genetically determined through the efficacy of repair but not in the manner of a strict programme for self-destruction. The question is not *whether* ageing is genetically controlled but, rather, *how* this is arranged and it seems constructive to limit the term 'programme' to mechanisms which specifically terminate the life span. This conceptual distinction is of major importance in relating cellular and molecular mechanisms of ageing to their broader biological role [12] (see below).

In species with life histories other than iteroparous ones, the term 'ageing' itself loses clarity [12]. For instance, in semelparous species, death and reproduction are usually closely linked and the mortality-inducing events associated with maturation are of fundamentally different biological significance from those associated with senescence [11, 12]. Detailed consideration of other life history patterns is beyond the scope of this brief review, but it may be noted that the principles which underlie the disposable soma theory can be extended to provide a broad framework within which a wide variety of basic questions about failure of life maintenance can be studied.

Organisms and Their Cells

The first theory of cellular ageing was formulated before animal cells could be grown in isolation. When tissue culture techniques were developed, however, *Carrel* and *Ebeling* claimed that isolated cells grew indefinitely [for review, see 37]. For almost 50 years this claim was accepted, and theories of ageing turned to external agents or to effects at the supracellular level, until *Hayflick and Moorhead* [4] demonstrated that normal human fibroblasts exhibit only finite growth in vitro. Since then, numerous laboratories have confirmed this observation and extended it to other vertebrate species.

It is now widely accepted that the limited replication of normal fibroblasts in tissue culture reflects ageing at the cellular level. Cultures grown from donors of increasing ages show, as predicted, a fall in division potential [20], while cultures established from different species show a positive correlation of in vitro life span with specific longevity [28]. Also, cells from patients with segmental progeroid syndromes, such as Werner's syndrome, have greatly reduced life spans [20]. Each of these findings strengthens the association of in vitro cell ageing with in vivo senescence, although evidence of a *causal* relationship is still lacking. This leaves room to argue that the loss of division potential in vitro is perhaps due to 'terminal' differentiation, not ageing [1, 16].

Evidence that several kinds of replicative cell exhibit only finite division in vivo has been obtained by serial transplantation in syngeneic mice. *Krohn* [17] showed that grafts of skin could be transferred successively from an old animal to a young one for a total period longer than the individual lifetime, but that the life span of the grafts was eventually limited. Similarly, serial transfer of mouse mammary gland tissue [2], haematopoietic cells [24] and antibody-forming cells [36] has been carried out through several passages, but not indefinitely.

The question arises naturally, if cellular ageing is an important determinant of life span, which types of cell are primarily responsible? No answer to this question is currently available. The fact that replicative cells cease dividing suggests that the inability to replenish them forever may be important. But, both in vivo and in vitro studies indicate that the capacity for cell renewal considerably exceeds normal lifetime demand. It is, therefore, likely that any in vivo relevance of replicative cell ageing is associated more with a reduction in the rate of cell renewal, coupled with an increasing cell failure rate, than with the ultimate cessation of growth. This probably occurs in parallel with the loss or malfunction of post-replicative cells, such as neurones, and the appearance of proliferative imbalance in cell populations which are continually turned over, such as may occur in autoimmune disease.

There is good reason, on both epidemiological and on evolutionary grounds, to expect ageing in vivo to be, on average, synchronous within different organs and tissues. Consistent failure of one system before others would generate conspicuous pathology, which is not the case [19], and would be a candidate for counter-selection. This suggests that the effect of ageing on cell proliferation should vary from one type of tissue to another, and possibly also from site to site, according to the role and normal growth kinetics of the cell population in vivo. It is found, for instance, that adult human glia-like cells enter a phase of proliferative decline after approximately 15 population doublings in vitro [27], whereas adult skin fibroblasts divide usually twice as much [3, 20], and that lung fibroblasts divide further, on average, than skin fibroblasts taken from the same fetus [31].

On the whole, the link between cellular and organismic ageing is the least well developed of the three components required for a complete theory of cell ageing, and progress in this area is a prerequisite for a fuller understanding. To develop a comprehensive link between cellular ageing and the senescence of the organism as a whole requires that the growth kinetics of cell populations in vivo, their interactions with each other, and the alterations of these parameters with age should be analysed in detail [7]. The most promising approach to this daunting task is likely to be through the construction of chimaeric animals, using the newly developed techniques of mammalian gene transfer [18].

Mechanisms of Cell Ageing

Two main kinds of mechanism for cellular ageing have been proposed, 'programme' theories and 'damage' theories. It is now apparent, for fibroblasts at least, that there is a major stochastic element in the senescent behaviour of individual cells [32]. However, this does not, by itself, provide evidence for the damage theories, since initiation of a genetic programme may also be probabilistic.

Among programme theories two sub-categories exist. In the first, a programme for cell ageing is regarded as the cause of ageing in the organism as a whole. Such theories face the objection, discussed previously, that it is exceedingly hard to explain *why* such a programme should have evolved. The second sub-category deals specifically with replicative cells and suggests that their limited growth results from programmed terminal differentiation into a non-proliferative state. It is readily seen that this group of theories in fact sidesteps the ultimate issue of cellular ageing, which is transferred to considering why terminally differentiated cells should be unable to function indefinitely and why the population of stem cells, from which the programmed cells arise, either senesces or is lost [24, 36]. This leads straight back to the problem of how age-related changes in the kinetics of growth of cell populations is related to senescence of the organism as a whole.

Damage theories of cellular ageing are multiple but share the unifying principle that slow accretion of random damage will result ultimately in loss of function in so large a proportion of replicative and post-replicative cells that the organism dies. The disposable soma theory predicts that for somatic cells the levels of repair will be insufficient to prevent such an accumulation and, thus, lends strong support to the damage theories, as a class. The theory also predicts that the levels of repair will be found to correlate with species' life spans and this, in principle, is readily testable. By comparative study of the efficiency of somatic repair systems in different species it may be possible to identify which of these systems are primarily responsible for regulating the duration of cellular and organismic life spans.

Reasoning similar to that used for senescent changes at the supracellular level suggests that loss of cell viability will be due to a broad spectrum of effects, rather than to a single specific cause. Thus, in the terminology of *Sacher* [30], single *aspect theories* based on individual features of the phenomenology of cell ageing are likely each to be true in part, but to fall short of providing a complete explanation. In the search for a primary cause of cell ageing, attention must be directed at the most fundamental level.

The common denominator of all cellular functions is the molecular machinery for replicating, repairing and decoding genetic information. These enzymatic processes operate with exquisite fidelity, but mistakes do nevertheless occur. The cost of proof-reading each step of DNA, RNA and protein synthesis is considerable and some trade-off between cost and accuracy seems likely [15]. Errors in DNA and in protein synthesis have been the basis of two distinct theories of cell ageing. The somatic mutation theory attributes cell ageing to the gradual accumulation of multiple defects in DNA [3, 33]. The protein error theory attributes cell ageing to the cyclic propagation of errors in the cellular transcription and translation apparatus, which can be independent of gene mutations [25]. In fact, on theoretical grounds, it appears hard to account for known features of cell ageing in terms of somatic mutations alone [6], and there is good reason to believe that somatic mutation should be considered as part of a more general error theory in which errors in all pathways of intracellular information transfer can participate and interact [6, 9, 26]. Such a theory, if correct, could account for each specific aspect theory, since breakdown of information transfer would result in progressive impairment of all cellular functions.

Evidence to test the general error theory remains at present inconclusive, partly because the theory's predictions are more complex than was originally recognised [9, 29], but particularly because experimental methods to detect and measure low frequency errors need to be further refined [29]. These techniques should be developed as a high priority, not only for their importance in studies of cell ageing, but also for their relevance to the whole field of accuracy in molecular biology. If error propagation turns out to play no part in cellular ageing, useful knowledge will in any case have been gained. If, on the other hand, error propagation is found to be of primary importance, a mechanism will have been discovered which can account simultaneously for the ageing of replicative and post-replicative cells.

Immortal Cells

The fact that heteroploid transformed cells, derived either from malignant tumours (e.g. the HeLa cell line) or by establishment in vitro, can grow indefinitely presents a serious challenge to theories of cell ageing. Any satisfactory mechanism to explain cellular ageing must be complemented by a description of how the mechanism may be altered or inactivated to allow indefinite growth. In the case of programme theories, this is easy. Interruption of the programme by mutation or epigenetic change is all that is required. In the case of damage theories the explanation is more complex and two alternatives can be considered.

Firstly, it is conceivable that there exist within the genome instructions for special error-correcting mechanisms that normally function only in the germ line. Such mechanisms would protect germ cells from most (but not necessarily all) of the damage which afflicts somatic cells but would be switched off for energy-saving reasons within the soma. Immortalisation of a clone of somatic cells would result if these mechanisms were accidentally re-activated by mutation or epigenetic change in gene expression, and this might account for the neoplastic reversion to a quasi-embryonic state which is so common in malignant cells.

The second route to immortalisation of somatic cells is through cellular selection. *Orgel* [26] made the simple, but profound observation that a population of cells in which each individual leaves, on average, less than one daughter will die out, while a population in which each individual leaves more than one daughter will multiply indefinitely. Thus, a minor change in the reproductive kinetics of individual cells can dramatically alter the behaviour of a population. A more elaborate version of this principle, and one which is better able to explain the growth properties of normal and transformed fibroblasts in vitro, is embodied in the commitment theory of cell ageing [5, 13]. In essence, this theory suggests that cells become committed at random to the process of senescence (e.g. by acquiring a critical level of damage) and that cellular selection can only compete effectively with this process if the deleterious effects of senescence are felt sufficiently soon. Accidental release of a cell from the normal internal checks on its proliferation, which presumably relate to its function in vivo, could thus intensify selection to a point where the least damaged (= least senescent) cells are continually preserved [5].

Either of these mechanisms could also explain the difference between mortal populations of somatic cells and the immortal lineage of germ cells. It is almost certainly the case that individual germ cells deteriorate with age, as shown by the increase in birth defects with maternal and paternal age, but, in the long run, the germ line must escape deleterious change. Of interest, also, in relation to cell ageing and germ line immortality are malignant teratocarcinoma cells, which resemble undifferentiated embryo cells in morphology and which grow indefinitely in culture, but which differentiate spontaneously into a wide variety of seemingly normal somatic cells of finite life span [21].

An important conclusion emerging from these considerations is that mortal and immortal cell populations may differ less in their *average* properties than the marked difference in their growth behaviour would suggest. Many of the individual cells in an immortal population may be undergoing an ageing process similar to that in the mortal population, with the immortal phenotype being sustained only by a small sub-population of cells. Thus, comparisons based on extracts from entire cell populations may fail to reveal the essential differences which exist among individual clones. To understand the nature of cell transformation fully, novel techniques may be required that are capable of detecting changes which occur in individual cell lineages.

Conclusions

Ageing belongs, like morphogenesis and cancer, to that challenging class of biological processes where the end result is obvious, but the mechanism remains obstinately obscure. The phenomenology of senescence is so widespread and diverse that there is an abundance of systems to study and it would be possible, with only observational studies, to sustain a prolific literature for many years to come. The concern is, however, that without a unified theoretical framework, to which new experimental results can be referred, progress in understanding what really causes senescence may be slow and inefficient.

Of the three components needed to construct a unified theory of cellular ageing, each is at a different stage of development. The first, that which seeks to explain the role of ageing in the organism as a whole, is the oldest and the most nearly complete. The greatest single contribution of this approach has been the dismissal of adaptive programme theories, except under very special circumstances. This calls for a re-orientation of much of gerontological thinking, which has yet to take place. The disposable soma theory, which embraces a number of earlier views [12], defines a strong new predictive basis for understanding the nature and evolution of senescence in iteroparous species, and further refinements are to be expected as a more detailed theory of life history optimisation takes shape. In particular, it emerges from the theory that great care should be exercised in extending the concept of ageing to species which reproduce only once in their lifetime, or which lack a clear distinction between soma and germ line [11].

The second component of the unified theory is the least developed of the three. This is the link between cellular and organismic ageing. It is tempting to speculate that, had *Carrel* and *Ebeling's* view on cellular immortality not been so widely accepted, this relationship would have received intensive study in the period before the flourishing of molecular biology made it possible to probe deeper inside the cell for the mechanism of cell ageing. In any event, the consequence is that the relevance of model systems for the study of cell ageing needs considerable further study, especially if the roles of ageing in replicative and non-replicative cells are to be united in a single theory.

Finally, the component which receives the most attention at present is the mechanism of cell ageing. Here the diversity of research is so great as to be almost bewildering, and unrelated pursuit of separate aspect theories is commonplace. While the heterogeneity of age-related changes in cells provides some justification for this fragmented assault, progress is likely to occur faster, and more usefully, if research is linked through a unified theory of cell ageing to conceptual developments taking place at the two higher levels of biological organisation.

References

1 Bell, E.; Marek, L.F.; Levinstone, D.S.; Merrill, C.; Sher, S.; Young, I.T.; Eden, M.: Loss of division potential in vitro: aging or differentiation? Science *202:* 1158–1163 (1978).

2   Daniel, C.W.: Finite growth of mouse mammary gland serially propagated in vivo. Experientia 29: 1422–1424 (1973).
3   Hayflick, L.: The limited in vitro lifetime of human diploid cell strains. Expl Cell Res. 37: 614–636 (1965).
4   Hayflick, L.; Moorhead, P.S.: The serial cultivation of human diploid cell strains. Expl Cell Res. 25: 585–621 (1961).
5   Holliday, R.: Growth and death of diploid and transformed human fibroblasts. Fed. Proc. 34: 51–55 (1974).
6   Holliday, R.; Kirkwood, T.B.L.: Predictions of the somatic mutation and mortalisation theories of cellular ageing are contrary to experimental observations. J. theor. Biol. 93: 627–642 (1981).
7   Kay, H.E.M.: How many cell generations? Lancet ii: 418–419 (1965).
8   Kirkwood, T.B.L.: Evolution of ageing. Nature, Lond. 270: 301–304 (1977).
9   Kirkwood, T.B.L.: Error propagation in intracellular information transfer. J. theor. Biol. 82: 363–382 (1980).
10 Kirkwood, T.B.L.: Repair and its evolution: survival versus reproduction; in Townsend, Calow, Physiological ecology, pp. 165–189 (Blackwell, Oxford 1981).
11 Kirkwood, T.B.L.: Comparative and evolutionary aspects of longevity; in Finch, Schneider, Handbook of the biology of aging; 2nd ed. (Van Nostrand & Reinhold, New York, in press, 1983).
12 Kirkwood, T.B.L.; Cremer, T.: Cytogerontology since 1881: a reappraisal of August Weismann and a review of modern progress. Hum. Genet. 60: 101–121 (1982).
13 Kirkwood, T.B.L.; Holliday, R.: Commitment to senescence: a model for the finite and infinite growth of diploid and transformed human fibroblasts in culture. J. theor. Biol. 53: 481–496 (1975).
14 Kirkwood, T.B.L.; Holliday, R.: The evolution of ageing and longevity. Proc. R. Soc. Lond. B 205: 531–546 (1979).
15 Kirkwood, T.B.L.; Holliday, R.: Selection for optimal accuracy and the evolution of ageing; in Galas, Accuracy in molecular biology (Dekker, New York, in press, 1983).
16 Kontermann, K.; Bayreuther, K.: The cellular ageing of rat fibroblasts in vitro is a differentiation process. Gerontology 25: 261–274 (1979).
17 Krohn, P.L.: Review lectures on senescence. II. Heterochronic transplantation in the study of ageing. Proc. R. Soc. Lond. B 157: 128–147 (1962).
18 Martin, G.M.: Cellular aging – postreplicative cells. Am. J. Path. 89: 513–530 (1977).
19 Martin, G.M.: Genetic and evolutionary aspects of aging. Fed. Proc. 38: 1962–1967 (1979).
20 Martin, G.M.; Sprague, C.A.; Epstein, C.J.: Replicative life span of cultivated human cells: effect of donor's age, tissue and genotype. Lab. Invest. 23: 86–92 (1970).
21 Martin, G.R.: Teratocarcinomas and mammalian embryogenesis. Science 209: 768–776 (1980).
22 Maynard Smith, J.: Group selection. Q. Rev. Biol. 51: 277–283 (1976).
23 Medawar, P.B.: An unsolved problem in biology (Lewis, London 1952).
24 Ogden, D.A.; Micklem, H.S.: The fate of serially transplanted bone marrow cell populations from young and old donors. Transplantation 22: 287–293 (1976).
25 Orgel, L.E.: The maintenance of the accuracy of protein synthesis and its relevance to ageing. Proc. natn. Acad. Sci. USA 49: 517–521 (1963).
26 Orgel, L.E.: Ageing of clones of mammalian cells. Nature, Lond. 243: 441–445 (1973).

27 Ponten, J.; Westermark, B.: Cell generation and ageing of non-transformed glial cells from adult humans. Adv. Cell Neurobiol. *1:* 209–227 (1980).
28 Rohme, D.: Evidence for a relationship between longevity of mammalian species and life spans of normal fibroblasts in vitro and erythrocytes in vivo. Proc. natn. Acad. Sci. USA *78:* 5009–5013 (1981).
29 Rosenberger, R.; Kirkwood, T.B.L.: The stability of the translation apparatus; in Galas, Accuracy in molecular biology (Dekker, New York, in press, 1983).
30 Sacher, G.A.: Theory in gerontology. Part I. Annu. Rev. Gerontol. Geriat. *1:* 3–25 (1980).
31 Schneider, E.L.; Mitsui, Y.; Au, K.S.; Stuart Shorr, S.: Tissue-specific differences in cultured human diploid fibroblasts. Expl Cell Res. *108:* 1–6 (1977).
32 Smith, J.R.; Whitney, R.G.: Intraclonal variation in proliferative potential of human diploid fibroblasts: stochastic mechanism for cellular aging. Science *207:* 82–84 (1980).
33 Szilard, L.: On the nature of the aging process. Proc. natn. Acad. Sci. USA *45:* 30–45 (1959).
34 Weismann, A.: Essays upon heredity and kindred biological problems; 2nd ed., vol. 1 (Clarendon Press, Oxford 1891).
35 Williams, G.C.: Pleiotropy, natural selection and the evolution of senescence. Evolution *11:* 398–411 (1957).
36 Williamson, A.R.; Askonas, B.A.: Senescence of an antibody-forming clone. Nature, Lond. *238:* 337–339 (1972).
37 Witkowski, J.A.: Dr. Carrel's immortal cells. Med. Hist. *24:* 129–142 (1980).
38 Woolhouse, H.W.: The nature of senescence in plants; in Woolhouse, Aspects of the biology of ageing, pp. 179–213 (Cambridge University Press, Cambridge 1967).

T.B.L. Kirkwood, PhD, National Institute for Medical Research, Mill Hill, London NW7 1AA (England)

Do Hemopoietic Stem Cells Age?

David E. Harrison

The Jackson Laboratory, Bar Harbor, Me., USA

Background

The rate of mammalian ageing is strictly species-specific. A mouse ages 30 times more rapidly than a human being. Therefore, the process by which the ageing rate is controlled is of central importance. In studies of cellular ageing, it is essential to learn whether ageing is intrinsically timed within the cells or tissues of interest. Studies of tissues transplanted from an old donor into a young recipient can be extremely useful in answering this question. Work in my laboratory has focused on hemopoietic stem cell lines. These cell types have extensive proliferative capacities and can be studied under natural conditions by transplantation in vivo. Figure 1 illustrates the type of experiment used initially. By transplanting tissue from an old donor into a young recipient, we could isolate the tissue from extrinsic factors causing ageing in the old donor.

For practical reasons, we define ageing as a loss of cell function with time. A tissue from an old donor that is able to function as well as the same tissue from a young donor is not intrinsically aged. Using this definition, transplantation experiments (fig. 1) answer a basic question about the ageing process: Are tissues able to function beyond the normal life span of the species? If not, the findings are consistent with a model proposing that ageing is intrinsic. If tissues can function beyond their normal life spans, the findings are consistent with a model proposing that ageing is extrinsic. The ageing process in such tissues must be timed to occur at a reduced rate; it may not occur at all. Obviously studies of such old donor tissues in young recipients may lead to important advances in the biology of ageing.

Although transplantation experiments (fig. 1) are basically simple, they have several pitfalls. The following four criteria [*Harrison*, 1973, 1978, 1983a,

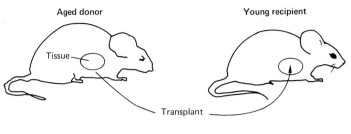

Possibility	Result
A. Ageing intrinsic in all cells	Defect continues
B. Ageing timed by one crucial tissue	Defect cured
C. Ageing results from interaction	Defect cured

Fig. 1. To determine whether or not ageing in a particular tissue is intrinsically timed, that tissue is transplanted from the aged donor into a young histocompatible recipient. If ageing is intrinsic (possibility A), the tissue will become defective with age, just as it would have had it been left in the old donor. If ageing is not intrinsic (possibility B except for the key tissue, possibility C for all tissues), the old tissue will not become defective but will function as well as transplanted young tissue [*Harrison*, 1978].

b] are useful in designing and interpreting transplantation experiments to avoid these pitfalls:

(1) Function. Ageing is defined as a loss of functional capacity. Therefore, tests that rigorously measure the function of the specific tissue transplanted are required.

(2) Identification. The transplanted tissue performing the measured functions must be unambiguously identified. For example, in an irradiated recipient regenerating host marrow and donor marrow must be distinguished. This is best done by using genetic markers.

(3) Control. Since transplantation itself may affect tissue function, tissues from young donors must be compared to those from old donors.

(4) Health. The old tissue must not be damaged irreversibly by its residence in the old donor. For example, marrow cell grafts must not include tumor cells.

Previous studies of hemopoietic stem cell transplantations have been reviewed elsewhere [*Harrison*, 1979a]. The most common pitfall encountered is the failure to meet criterion 2. In order to distinguish cells produced by donor and host stem cells, a wide variety of genetic markers can be used. These are discussed in the review.

Do Hemopoietic Stem Cells Age? 23

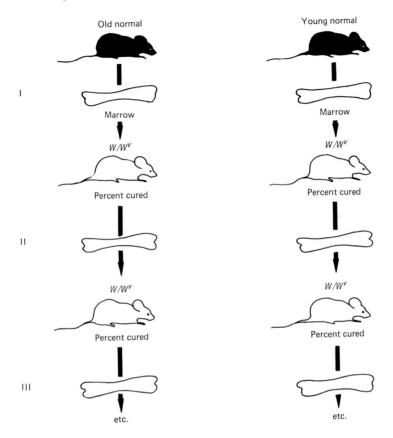

Fig. 2. Stem cells from old and young normal donors are compared in parallel experiments to distinguish the effects of ageing from effects of transplantation [*Harrison*, 1978]. Experimental details are given in table I.

Design of Initial Experiments

In initial experiments, unirradiated genetically anemic W/W^v mice were used as recipients. They have a macrocytic anemia that is permanently cured by grafting histocompatible normal (+/+) marrow cells. Erythrocytes produced by the cured W/W^v mice have the hemoglobin types of the donors, showing that the donor stem cells have repopulated the W/W^v recipients [*Russell and Bernstein*, 1968]. Figure 2 illustrates the experimental design used to test the function of old hemopoietic stem cells in young W/W^v recipi-

Table I. Effects of donor age and numbers of transplantations on hemopoietic function of stem cells

Original donor line		Percentage W/W^v cured		Number of CFU-S per 10^5 cells
age	number of serial transplants	1–3 months	6–9 months	
Old	1	95 ± 4(20)	82 ± 8(15)	9.1 ± 1.4(7)
Young	1	86 ± 7(22)	84 ± 8(17)	9.9 ± 1.3(6)
Old	2	88 ± 5(23)	58 ± 8(16)	5.3 ± 1.4(10)
Young	2	94 ± 2(22)	69 ± 8(17)	7.8 ± 1.8(9)
Old	3	69 ± 8(23)	52 ± 7(25)	6.2 ± 0.6(15)
Young	3	83 ± 5(28)	68 ± 8(29)	7.5 ± 0.8(19)

Data are given as mean ± SE (number of donors used). An average of 5 (range 2–11) WBB6F$_1$-W/W^v recipients were used with each donor. Numbers of CFU-S (macroscopic spleen colonies) are given per 10^5 donor marrow cells; generally 4–6 irradiated recipients were scored for each donor. The donor genotypes were WCB6F$_1$ or WBB6F$_1$; these mice are similar so the results were pooled. Transplantations were performed at intervals of 300–550 days, and 3–10×10^6 (usually 10×10^6) marrow cells were given to each W/W^v recipient. The same stem cell lines were serially transplanted and tested; only cured recipients were used as donors for the next successive transplantation. The ages of the original donors (given as mean and range) were 33.6 (29–39 months) for old and 6.1 (1–12) months for young donors. Most of these data have been published [*Harrison, 1979b*].

ents. After the first transplantation, the numbers of W/W^v recipients cured by equal numbers of bone marrow cells from old and young donors were similar (transplantation 1, table I). Most of the W/W^v recipients remained cured, and macroscopic spleen colony-forming cells were found in their marrow. Such cells are not found in W/W^v mice. To test the function of old stem cells more rigorously, their descendants were serially transplanted at annual intervals as shown in figure 2. The interval between transplants was long and cell numbers were high (usually 10 million marrow cells per recipient) to minimize the effects of serial transplantation. Results from many such experiments are displayed in table I. During the first three serial transplantations, stem cell lines from old and young donors functioned at similar levels (table I) [*Harrison, 1972, 1973*].

Other tests besides curing genetically anemic recipients showed that the old stem cells continue to function for periods of time well beyond the

Do Hemopoietic Stem Cells Age?

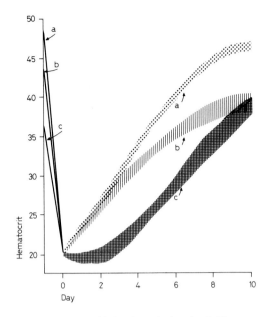

Fig. 3. Mice were bled 3 times during day 0. The recovery curves for each of the groups were obtained by regression analysis and are plotted as patterns showing 95% confidence limits. Pattern 'a' defines the responses of 5- to 8-month-old B6 and WBB6F$_1$ normal mice tested with and without young marrow cell injections, and of 5- to 10-month-old W/W^v mice cured with B6 or WCB6F$_1$ marrow cells 3–8 months before bleeding. The marrow cells had been serially transplanted 2–4 times and had functioned normally for 18–73 months. Pattern 'b' defines the responses of 25-month-old B6 mice given young marrow cell injections 8 days and 1 day, or 1 day before bleeding. It also includes 25-month-old B6 mice that did not receive marrow cell injections, since their responses were the same. Pattern 'c' defines the responses of uncured W/W^v mice [from *Harrison*, 1975b].

maximum life span of the mouse species used *(Mus musculus).* Genetically anemic mice, or aged mice, recover from severe bleeding more slowly than normal; however, genetically anemic mice cured by old marrow cells recover as rapidly as those cured by young marrow cells or normal mice [*Harrison*, 1973, 1975b]. Therefore, the defects causing aged mice to recover slowly are not intrinsic in the stem cells. The results of many such experiments are summarized in figure 3. Note that genetically anemic mice (pattern c) require several extra days before they begin to recover from bleeding, while old mice (pattern b) begin to recover at normal rates, but then are not able to sustain the rapid rate of recovery. Anemic mice cured with old or young cells show

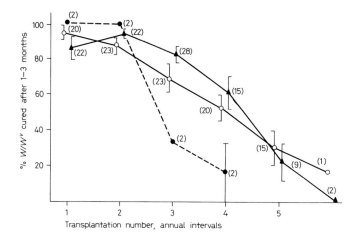

Fig. 4. Marrow cell lines from WCB6F$_1$ donors were serially transplanted annually into young W/W^v recipients. Young controls ($\triangle$) had a mean age 6.1 months (range 1–12 months). Old donors ($\bigcirc$) had a mean age of 33.6 months (range 29–39 months). 6-month-old donor cells ($\bullet$) were transplanted every 3 months rather than annually. Donor ages are those at the initial transplantation. The number of donors used at each transplantation is in parentheses, with a mean of 5.1 (range 2–11) young W/W^v recipients per donor. Each point on the figure is the mean for all the donors it represents, and brackets indicate 1 SE [from *Harrison,* 1979b].

pattern a, the same as young normal (+/+) mice. This was found even in anemic mice cured by stem cell lines that had been serially transplanted up to four times and were 73 months old.

Numbers of Serial Transplantations

Since marrow stem cells from old donors functioned normally, they were tested by serial transplantation at annual intervals until they failed to cure W/W^v recipients [*Harrison,* 1975a, 1979a]. Figure 4 summarizes results of these experiments. At the third or fourth serial transplantation, the percent of W/W^v recipients cured begins to decline significantly. This decline occurs despite the facts that (1) large numbers of marrow cells were transplanted, and (2) long intervals were allowed to occur between stem cell transplantations. There are only a few stem cell lines that continue to function after the fifth serial transplant, and none are functional after the sixth serial transplant (fig. 4). Although there was a tendency for the percentage cured to decline

Table II. Life spans of longest-lived hemopoietic stem cell lines

Donor strain	Number of serial transplants	Age, days	% W/W^v cured
B6	4–5	2,593–3,044	86 ± 9 (4)
WCB6F$_1$	5	2,679–2,687	67 (3)
WBB6F$_1$	4	2,818–2,821	100 (2)

The recipients were W/W^v mice. Cells were serially transplanted at 300- to 550-day intervals from 29- to 39-month-old donors. The percent of W/W^v cured is given as mean ± SE (n) where n is the number of donors whose cells function normally (cured at least 2/3 of the W/W^v recipients) [*Harrison*, 1979b].

slightly more precipitously when cells from old rather than young donors were serially transplanted, these differences were not significant.

Even with limits on the number of successful serial transplants that they undergo, mouse hemopoietic stem cells often function for as long as 2,600–3,000 days (table II). Normal mean longevities of the mouse strains used as donors are 900–1,000 days, and maximum longevities are 1,200–1,300 days at best [*Myers*, 1978; *Harrison*, 1982]. The longest-lived mice of the *Mus musculus* species are F$_1$ hybrids in experiments involving life long caloric restriction with micronutrient supplementation. They appear to have maximum longevities of about 1,600 days [*Weindruch and Walford*, 1982, personal commun.]. Table II shows that stem cells are capable of curing W/W^v recipients for nearly twice this long. Stem cell life span is limited by the numbers of transplantations, not by age [*Harrison*, 1975a, 1979a]. If transplantation could be avoided, hemopoietic stem cells would probably function much longer, as will be discussed later in this report. However, we can already see that ageing in hemopoietic stem cells is not nearly as rapid as ageing in the tissues that limit the life span of a mouse.

This conclusion is supported by a variety of stem cell transplantation experiments summarized by *Harrison* [1975a, 1979b]. Mice of the WBB6F$_1$, WCB6F$_1$ and B6 genotypes were studied for five to six serial transplantations, while several other genotypes were studied over two serial transplantations. These experiments lead to a general conclusion: as long as old donors do not have diseases that affect their marrow, their stem cells function as well as those from young donors. The loss in function that begins after three or four serial transplantations appears consistently both in our experiments and in those of other workers [reviewed in *Harrison*, 1979a, 1983a, b]. Serial transplantation, not ageing, is responsible for this loss of stem cell function.

Stem Cells of the Immune System

Changes in function that occur during ageing in stem cells that repopulate the immune system are a source of current controversy. In short-term experiments, old stem cells may not have time to recover from the deleterious effects of residence in the old donor. This may explain why marrow cells from old BC3F$_1$ mice produce half as many B lymphocytes as marrow from young donors after 4 days in irradiated recipients [*Farrar* et al., 1974]. It may also explain why marrow cells from old BC3F$_1$ mice recovered more slowly than did marrow cells from young BC3F$_1$ mice following two doses of sublethal irradiation 5 h apart [*Chen*, 1974]. Even in long-term experiments, some workers have reported defects in function of stem cells from old animals. *Albright and Makinodan* [1976] found that stem cells from old mice grew more slowly than those from young donors. When spleen cells from old BC3F$_1$ mice were transplanted into young recipients by *Price and Makinodan* [1972], immune responses fell to the level seen in the old mice by 60 days. Similar results were found by *Kishimoto* et al. [1973, 1976]. They reported that spleen cells from irradiated recipients populated by old B6 marrow grafts failed to induce graft-versus-host reactions 80–100 days after grafting, while recipients of grafts from 2-month-old donors gave better than normal graft-versus-host reactions. 4–6 weeks after transplantation, marrow grafts from 15- to 18-month-old B6 mice had not recovered normal B cell functions.

We investigated stem cells of the immune system because current theories suggest that precursors for hemopoietic and immune functions are the same (fig. 5). Our initial results suggested that stem cells from old donors repopulated the immune system as well as did those from young donors. We compared immune responses in irradiated recipients repopulated by marrow or spleen cell transplants from old and young donors. The recipients were indistinguishable in their ability to produce direct plaques when stimulated with sheep erythrocytes. 80–100% of the mitotic cells in their lymph nodes, marrows, thymuses and Peyer's patches had the donor-type chromosome marker [*Harrison and Doubleday*, 1975]. Our later results confirmed these findings in experiments using old and young donors whose immune responses were measured. Marrow cells from old donors with defective

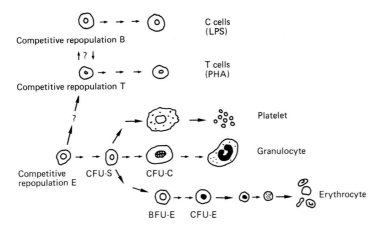

Fig. 5. Stem cell differentiation in the immunohemopoietic systems may involve a single cell type, or several cell types, that multiply to replenish themselves and differentiate to populate the animal. The newly developed competitive repopulation assay was designed to measure the functional abilities of these earliest precursors. Competitive repopulation B, T, and E assays, respectively, test precursors of: B cells proliferating in response to lipopolysaccharide (LPS) in vitro, T cells proliferating in response to PHA in vitro, and erythropoietic cells producing hemoglobin, or proliferating in response to severe bleeding in vivo. In mitotic cells chromosome markers are identified as detailed in *Harrison* et al. [1978]. Competitive repopulation E assays are best performed using hemoglobin markers [*Harrison*, 1980]. Cells that form macroscopic colonies on spleens of irradiated mice (CFU-S) are the earliest precursors detectable as single cells; they differentiate to form platelets, granulocytes, and erythrocytes. Colony-forming units in semisolid agar cultures (CFU-C) produce granulocytes. Burst-forming units and colony-forming units that respond to erythropoietin in vitro (BFU-E and CFU-E) are, respectively, earlier and later precursor cells committed to form erythrocytes [from *Harrison*, 1984].

immune responses support similar levels of immune responses in young recipients as do marrow cells from young controls [*Harrison* et al., 1977].

Other data also suggest that old immune stem cells are not defective. After hypophysectomy, the thymus weights and cortical areas increase substantially, and certain thymus-dependent immune responses are rejuvenated in old mice [*Harrison* et al., 1982]. Such improvements would not be seen if the defects that occur with ageing were intrinsic to the stem cells of old mice. The results of our experiments seem to contradict previous reports implying that immunological defects with age are intrinsic to stem cells. Perhaps some of the old donors of the $BC3F_1$ and B6 genotypes used in these studies contained tumor precursor cells that were inadvertently transplanted with the marrow or spleen cells. Such cells could severely suppress immune responses even when present in very small proportions [*Jaroslaw* et al., 1975].

Another reason for the contradiction is suggested by recent evidence on changes that occur in regulatory cell types with age. *Tyan* [1980] found changes with age in both helper and suppressor T cells that have direct effects on hemopoietic precursor cells. B cell function can be affected by a decline with age in the T cells necessary for B cell maturation [*Szewczuk* et al., 1980]. Suppressor T cells are more easily induced in old than in young mice, but they suppress the antibody response against azobenzenearsonate less efficiently in cells from old mice [*Doria* et al., 1982]. Although old stem cells support normal responses to the stress of endotoxin and chronic bleeding [*Tyan*, 1982a], they fail to generate normal numbers of lymphocytes or normal levels of hemoglobin [*Tyan*, 1982b]. In recipients of old marrow, cellular immune responses become deficient after 8 months, while hemopoietic cell production remains normal in the same individuals [*Gozes* et al., 1982]. These reports suggest that age-dependent regulatory cell changes may only affect certain stem cell functions.

Competitive Repopulation Assay

The regulatory cells that appear to change with age may be included in marrow transplants. A new stem cell assay should be used to control for this possibility. The competitive repopulation assay compares repopulating abilities of two types of stem cells that are genetically distinguishable. The two types are mixed and injected into the same recipient. By using this assay, both types of stem cell are subjected to the same regulatory cell influences. Another advantage is that both short-term rapid repopulation and long-term maintenance of stem cell function are required for successful competition. Using this technique *Ogden and Micklem* [1976] analyzed recipients of mixtures of old and young marrow cells distinguished by a chromosome marker. In one case the progeny of the young donor cells predominated in the recipients, while in the other case the opposite was observed.

A difficulty with this technique is that only 2 donors can be compared at once. We have altered the technique to permit analysis of several old and young donors in the same experiment. To do this, we mix marrow from each of the old and young donors with a specific dose of genetically distinguishable marrow from a single pool [*Harrison* et al., 1978]. Cells from the pool are called competitors, and several recipients are given portions of each donor-

Table III. A mitotic competition experiment

Genotype	Marker	Use in experiment
CBA/H-T6J	2 T6 chromosomes	donor type: individual old or young donor cell lines
(CBA/H-T6J × CBA/CaF)F_1	1 T6 chromsome	competitor type: pooled young cells transplanted for the first time
CBA/CaJ	none	recipient type: lethally irradiated

Procedure: (1) Mix 3×10^6 marrow cells from each donor with 3×10^6 marrow cells of the competitor type. (2) Inject mixtures intravenously into recipients. (3) After 1–14 months determine the percentages of mitotic cells having 0, 1, or 2 T6 chromosome translocations after bleeding in vivo or PHA in vitro [*Harrison* et al., 1978].

competitor cell mixture. Cells from many different donors can be mixed with cells from the same competitor pool. The relative repopulating abilities of these donors are determined from the percentage of donor-type cells produced in the recipients, since all donors compete against the same competitor.

We wanted to define the repopulating abilities of specific cell types, such as erythroid cells, B cells and T cells. It is important to determine whether there is an age-dependent change in the capacity to produce one type but not others. Chromosome markers are analyzed after stimulating proliferation of a single cell type. They are only visualized in mitotic cells, so those responding to the cell type-specific stimulus are scored. Stimuli used include bleeding in vivo, and T and B cell mitogens in vitro [*Harrison* et al., 1978; *Harrison*, 1981]. Hemoglobin markers are convenient to use for measuring the competitive repopulating abilities of erythropoietic precursor cells [*Harrison*, 1980].

Table III illustrates the experimental design when chromosome markers are used. Old and young donors of a genotype having two of the T6 chromosome markers (CBA/H-T6J) are prepared. The marrow cells are mixed with a standard dose of pooled competitor cells from histocompatible mice having a single T6 chromosome [(CBA/H-T6J × CBA/CaJ)F_1 hybrids]. The mixtures are injected into lethally irradiated histocompatible recipients with no T6 chromosomes (CBA/CaJ). The percentage of cells containing 0, 1, or 2 T6 chromosomes is determined after severe bleeding to stimulate erythropoiesis in vivo, or phytohemagglutinin (PHA) treatment to stimulate T cells in

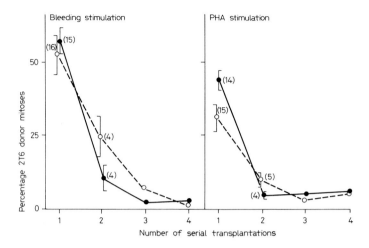

Fig. 6. Effects of age and serial transplantation on competitive repopulating ability. Values in parentheses are the number of donors, with 2.3 recipients per donor. Old donors (○) were 24- to 31-month-old and young donors (●) were 3- to 6-month-old CBA-T6 mice; this genotype has 2 T6 chromosomes. For transplantations 2, 3, and 4 their marrow was serially transplanted after 4–12 months. Competitive repopulation assays were performed as outlined in table III. Recipients were splenectomized after 1–14 months to provide spleen cells for PHA stimulation in vitro, and bled 1–4 weeks later to stimulate marrow cells in vivo. A mean of 48 (range 20–65) mitoses were scored for each determination. Usually cells from 2 old and 2 young donors were studied in each assay; in three experiments, donors at transplantation 2 were included, and in one experiment, donors at transplantations 3 and 4. Means for all donors are plotted, bars give ± SE [from *Harrison and Astle*, 1982].

vitro. In these experiments, erythropoietic and T cell types are assessed separately.

Figure 6 shows that results are similar for both cell types. Stem cells from old and young donors compete equally well against competitor cells from the same pool. However, after a single serial transplantation (transplantation 2), the repopulating ability of cells from carriers declines substantially, regardless of the age of the original donor [*Harrison* et al., 1978; *Harrison and Astle*, 1982]. The deleterious effects of transplantation have been confirmed by *Ross* et al. [1982].

The deleterious effects of transplantation may result from accelerating the mechanisms that cause normal ageing. If this is the case, the following comparison is valid. Compared to a life span of normal function, a single transplantation causes 3–7 times more decline in repopulating ability of PHA-responsive cells, and at least 10 times more decline in erythroid cells

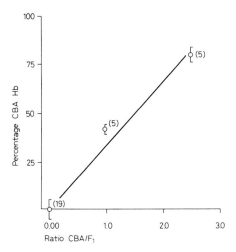

Fig. 7. The percent CBA hemoglobin (Hb) increased with the ratio of CBA/B6CBAF$_1$ marrow cells in the mixture injected into lethally irradiated B6CBAF$_1$ recipients. The correlation coefficient between the ratios and the percentages was 0.98 for this experiment in which 0, 10^6 or 2.5×10^6 CBA marrow cells were mixed with aliquots of 10^6 B6CBAF$_1$ marrow cells. Points are plotted for mean values, bars enclose ± SE, and numbers of recipients at each point are inside parentheses [from *Harrison*, 1980].

(fig. 6). Stem cell lines can be serially transplanted 4–6 times before losing their ability to repopulate and cure genetically anemic mice (fig. 4). Therefore, stem cells should be able to function normally through at least 15–50 life spans. On the other hand, the deleterious effects of transplantation may result from mechanisms independent of the ageing process. Even in this case, the similar functional abilities of stem cells from young and old donors suggest that they age very much more slowly than their donor [*Harrison and Astle*, 1982]. In fact, hemopoietic stem cells may not age intrinsically [*Harrison*, 1979a].

Competitive Repopulation Using Hemoglobin Markers

When hemoglobin markers are used rather than chromosome markers, the competitive repopulation assay measures erythropoietic stem cell function specifically. To use hemoglobin markers, electrophoretically separable hemoglobins must distinguish the donor and competitor types. Stem cells from each donor are mixed with a specific number of stem cells from a single

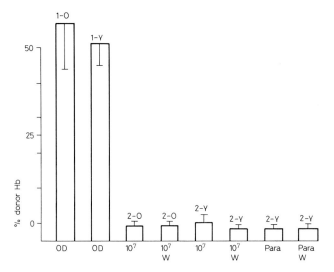

Fig. 8. Old (1-O) and young (1-Y) WBB6F$_1$ +/+ donors were 25–28 and 2–8 months old, respectively. Genetically anemic W/W^v (W) or lethally irradiated normal WBB6F$_1$ carriers were given donor marrow cells intravenously, and after 111 days were tested in the competitive repopulation assay; this was transplantation number 2 (2-O or 2-Y). Some donors were parabiosed (Para) to carriers for the 111 days. In the competitive repopulation assay, competitors (pooled marrow from 4- to 6-month-old donors) were B6 mice, recipients were WBB6F$_1$ mice, and mixtures of 3×10^6 donor cells with 7.5×10^6 competitor marrow cells were given to each recipient 113–211 days before its hemoglobins (Hb) were assayed. Data bars are given as mean ± SE. Numbers of recipients were 6–20 [from *Harrison and Astle*, 1982].

pool of competitors. Stem cell-depleted recipients are given portions of each donor-competitor cell mixture [*Harrison*, 1980]. Cells from a variety of different donors are mixed with cells from the same competitor pool. The relative repopulating abilities are determined by calculating the percentage of donor type hemoglobin produced in each of the recipients. Figure 7 shows that the percentage is proportional to the ratio of donor marrow cells.

When hemoglobin markers are used, stem cells from old donors compete at least as well as those from young donors. However, a single transplantation causes severe deleterious effects. The deficiencies are even greater than those shown when chromosome markers are used. When old and young WBB6F$_1$ marrow donors were competed against portions of young B6 marrow, the percentages of donor hemoglobin were 57 ± 13 and 51 ± 6% from 12 old and

Do Hemopoietic Stem Cells Age?

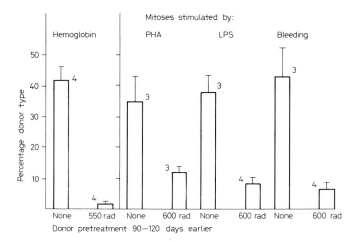

Fig. 9. Donors were tested in the competitive repopulation assay either without pretreatment or after receiving 550–600 rad gamma (cesium-137) irradiation 90–120 days previously. Hemoglobin markers were used to distinguish erythrocytes produced by stem cells from 8-month-old CBA-T6 donors and those from 5-month-old B6CBAF$_1$ competitors in the left-hand panel; mixtures of 1×10^6 donor and competitor marrow cells were given to each lethally irradiated B6CBA-T6F$_1$ recipient 90 days before hemoglobins were assayed. Chromosome markers were used to distinguish mitoses stimulated by PHA or lipopolysaccharide with spleen cells in vitro or by bleeding with marrow cell in vivo in the next three panels. Donors were 6-month-old B6CBA-T6F$_1$ mice while competitors were 5-month-old B6CBAF$_1$ mice. Each lethally irradiated B6CBAF$_1$ recipient was given 3×10^6 donor and competitor marrow cells 75–150 days before its cells were stimulated for the assay, and an average of 48 mitoses were scored per recipient. Data bars show mean ± SE; numbers of recipients are given by the top of each bar [from *Harrison*, 1983].

20 young donors, respectively. Several months after a single serial transplantation, neither old nor young WBB6F$_1$ stem cells produced detectable levels of hemoglobin when competed against the same portions of young B6 marrow cells (fig. 8). We tried to overcome the severe deleterious effect of transplantation by using W/W^v anemic recipients or by parabiosing the donor and recipient. Figure 8 shows that these procedures did not alter the effects of transplantation.

The competitive repopulating ability of stem cells is also severely reduced after recovery from sublethal irradiation (550–600 rad) (fig. 9). This dramatic effect of sublethal irradiation persists for several months and has only been detected using the competitive repopulation assay [*Harrison and Astle*, 1982; *Harrison*, 1983]. Transplantation of successively larger num-

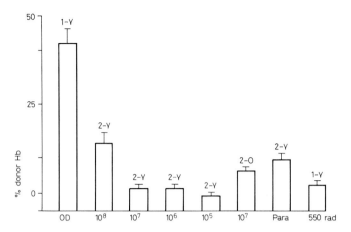

Fig. 10. The young (1-Y) CBA-T6 donor was 8 months old. For the second transplant (2-Y or 2-O) lethally irradiated B6CBA-T6F$_1$ carriers were given the donor marrow cell numbers listed intravenously, and after 90 days were tested in the competitive repopulation assay. Lethally irradiated CBA carriers in bar 7 received cells by parabiosis from the CBA-T6 donors for 90 days; CBA-T6 donors in bar 8 had been irradiated with 550 rad (gamma) 90 days before the assay. In the competitive repopulation assay, competitors (pooled marrow from 3.5- and 6.5-month-old donors) were B6CBAF$_1$ mice, recipients were B6CBA-T6F$_1$ mice, and mixtures of 10^6 donor and competitor marrow cells were given to each recipient 70–80 days before its hemoglobin was assayed. Data bars give mean ± SE. 4 recipients were used in each [from *Harrison and Astle*, 1982].

bers, 10^5, 10^6, or 10^7 marrow cells, into lethally irradiated carriers did not improve the competitive repopulating abilities of the stem cells from these carriers (fig. 10). Only in carriers receiving 10^8 marrow cells were competitive repopulating abilities improved, although they were still severely diminished when compared to the repopulating abilities of untransplanted marrow [*Harrison and Astle*, 1982].

The competitive repopulating technique is able to detect severe deleterious effects resulting from a single transplantation or from sublethal irradiation that are not detectable using other means of assaying stem cell function. Therefore, this technique is the most likely to detect small changes with age in stem cell function. The results suggest that old and young stem cells function equally well (fig. 6, 8). Therefore, within the limits of the most sensitive stem cell assays, hemopoietic stem cells are not affected by a life span of normal function.

Table IV. Effect on immunological enhancement and longevity of old marrow

Treatment	Age test days	Immune responses, % of young[a]		Life span days
		SRBC	PHA	
None	305 ± 10(29)	87.5 ± 4.3(26)	92.2 ± 4.8(21)	–
None	757 ± 10(20)	41.1 ± 6.3(10)[b]	30.0 ± 8.3(14)[b]	1,063 ± 31(20)
I, YM, T	766 ± 12(20)	81.1 ± 13.4(15)	77.5 ± 13.1(13)	968 ± 28(18)[b]
I, OM, T	770 ± 17(9)	24.8 ± 7.5(9)[b]	26.3 ± 9.7(7)[b]	904 ± 27(9)[b]

All data given as mean ± SE (number of values). I, YM, T treated mice had been lethally irradiated, repopulated with 1×10^7 marrow cells from syngeneic young adults, and given an infant thymus graft 4–8 months before immune responses were measured; I, OM, T treatment was the same, except that the marrow was from donors more than 750 days old. Mice were B6CBAF$_1$, WBB6F$_1$, or B6D2F$_1$ hybrids.

[a] Responses are expressed as percent of the responses of the highest responding young controls tested simultaneously. Responses to sheep red blood cells (SRBC) were measured as direct plaques in vivo, and responses to PHA were measured as ^{3}H-Thy incorporation in vitro.

[b] Significantly smaller than other results in the same column, $p < 0.05$ by the Student-Newman-Keuls multiple range test [*Astle and Harrison*, 1984].

Current Research Areas

An important question is whether changes that occur with age in stem cell immune functions are intrinsic to the stem cells themselves. We have repeated the experiments of *Hirokawa* et al. [1976, 1982] showing that old recipients are immunologically rejuvenated when given lethal irradiation, followed by grafts of young marrow and an infant thymus (table IV). Table IV also shows that such old recipients are not rejuvenated when given old marrow. Thus old marrow functions as well as young in restoring immune responses of young recipients [*Harrison and Doubleday*, 1975; *Harrison* et al., 1977], but fails to restore immune responses of old recipients.

Our data are consistent with the interpretation that old stem cells in the immune system develop intrinsic defects with age that are only expressed in old recipients. However, the normal function of old stem cells in the sensitive competitive repopulation assay makes this seem unlikely. An alternative explanation is suggested by the following experiment: When old and young histocompatible mice have their circulations joined by parabiosis, the immune responses of the old partners are not affected, but those of the young partners

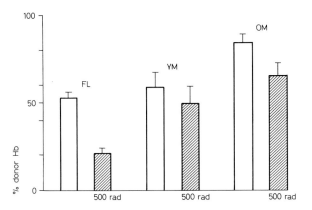

Fig. 11. Hemoglobin (Hb) percentage bars give mean ± SE. There were 7–8 donors with a mean of 3.4 recipients per donor. Liver (FL) was taken from 15- to 16-day fetuses; young marrow donors (YM) averaged 4.1 months and old marrow donors (OM) 26.4 months of age. All donors were B6 mice. Recipient hemoglobins were initially checked 80–100 days after male WBB6F$_1$ recipients were lethally irradiated and grafted with a mixture of 2.5×10^6 cells from the appropriate B6 donor plus 1.0×10^6 marrow cells pooled from 6-month-old WBB6F$_1$ +/+ competitors. Recipients were given an additional 500-rad gamma (cesium-137) irradiation, and hemoglobins were checked a second time 77 days later (187–207 days after mixtures were transplanted) [from *Harrison*, 1983].

are reduced to old levels [*Harrison*, 1979c; *Butenko and Gubrii*, 1980; *Astle and Harrison*, 1984]. Old animals appear to contain suppressor cells or factors that circulate into the young partners and suppress their immune responses. We hypothesize that the suppressors are produced by precursor cells that are found in old marrow. Furthermore, the precursors only produce suppressors in old recipients, although the suppressors are active if they circulate into young partners.

In another current experiment, the performances of stem cells from fetal liver, young adult marrow, and old adult marrow have been compared in competitive repopulation experiments using hemoglobin markers [*Harrison*, 1983]. C57BL/6J stem cells from old marrow, young marrow or fetal liver are mixed with competitor stem cells from young WBB6F$_1$ marrow. The hemoglobins produced by B6 and WBB6F$_1$ cells are quantitated after electrophoresis, and the amounts produced by each type are proportional to the numbers of marrow cells of that type in the mixture [*Harrison*, 1980]. Fetal cells are expected to show superior function as has been previously reported [*Micklem et al.*, 1972]. However, descendants of stem cells from old donors produced

25% more hemoglobin than those from young donors and 31% more than those from 15- to 16-day fetal donors (fig. 11).

It is possible that the stem cells from old donors have an initial advantage, due to changes in regulatory functions or to an accumulation of immune responsive cells in the marrow that could react against the competitors. To remove such effects, recipients of the stem cell mixtures were sublethally irradiated with 500 rad, a dose causing large effects on repopulating abilities, as shown in figure 9, and requiring an intense period of rapid growth to recover. Although the percentage of $WBB6F_1$ competitor cells increased in all cases, the stem cells descended from old marrow retained their advantage (fig. 11). Regardless of what caused the advantage, its persistence after many months and after sublethal irradiation suggests that there is no difference in functional ability between stem cells from old, young, or fetal donors [*Harrison*, 1983]. If this proves to be true, the answer to the question posed in the title of this paper is that stem cells do not age.

It may not be possible to establish absolute proof that stem cells, or any other somatic cell type, are immortal. However, the studies reviewed in this paper strongly suggest that possibility. Even the most sensitive tests of long-term stem cell function fail to establish functional losses with age in this cell type. Future theories of cellular ageing should consider this observation. Even by conservative interpretations, we have shown that ageing in stem cells is very much slower than in the systems limiting life expectancies of the stem cell donor. An exciting area for future analysis will be to explore the mechanisms by which stem cells minimize ageing rates and to determine whether these mechanisms can be extended to other cell types.

Acknowledgments

I thank *Clinton Astle* for collaboration and assistance throughout these experiments, Drs. *Elizabeth Russell* and *Jane Barker* for valuable discussions, and *Joan DeLaittre* and *Jack Doubleday* for dependable technical assistance. This work was supported by research grants AG-00594 and AG-01755 from the National Institute on Aging, and AM-25687 from the National Institute of Arthritis, Diabetes, Digestive and Kidney Diseases.

References

Albright, J.F.; Makinodan, T.: Decline in the growth potential of spleen-colonizing bone marrow cells of long-lived mice. J. exp. Med. *144:* 1204–1213 (1976).

Astle, C.M.; Harrison, D.E.: Beneficial effects of young but not old marrow on immune responses of aged mice: studies on the cellular level. J. Immun. *132* (in press, 1984).

Butenko, G.M.; Gubrii, I.B.: Inhibition of the immune responses of young adult CBA mice due to parabiosis with their old partners. Exp. Gerontol. *15:* 605–610 (1980).

Chen, M.G.: Impaired Elkind recovery in hemopoietic colony-forming cells of aged mice. Proc. Soc. exp. Biol. Med. *145:* 1181–1186 (1974).

Doria, G.; Mancini, C.; Adorini, L.: Immunoregulation in senescence: increased inducibility of antigen-specific suppressor T cells and loss of cell sensitivity to immunosuppression in aging mice. Proc. natn. Acad. Sci. USA *79:* 3803–3807 (1982).

Farrar, J.J.; Loughman, B.E.; Nordin, A.A.: Lymphopoietic potential of bone marrow cells from aged mice: comparison of the cellular constituents of bone marrow from young and aged mice. J. Immun. *112:* 1244–1249 (1974).

Gozes, Y.; Umiel, T.; Trainin, N.: Selective decline in differentiating capacity of immuno-hemopoietic stem cells with aging. Mech. Age. Dev. *18:* 251–259 (1982).

Harrison, D.E.: Normal function of transplanted mouse erythrocyte precursors for 21 months beyond donor lifespans. Nature new Biol. *237:* 220–222 (1972).

Harrison, D.E.: Normal production of erythrocytes by mouse marrow continuous for 73 months. Proc. natn. Acad. Sci. USA *70:* 3184–3188 (1973).

Harrison, D.E.: Normal function of transplanted marrow cell lines from aged mice. J. Geront. *30:* 279–285 (1975a).

Harrison, D.E.: Defective erythropoietic responses of aged mice not improved by young marrow. J. Geront. *30:* 286–288 (1975b).

Harrison, D.E.: Is limited cell proliferation the clock that times aging?; in Behnke, Finch, Moment, The biology of aging, pp. 33–35 (Plenum, New York 1978).

Harrison, D.E.: Proliferative capacity of erythropoietic stem cell lines and aging. An overview. Mech. Age. Dev. *9:* 409–426 (1979a).

Harrison, D.E.: Mouse erythropoietic stem cell lines function normally 100 months: loss related to number of transplantations. Mech. Age. Dev. *9:* 427–433 (1979b).

Harrison, D.E.: Treatments that retard or reverse immunological losses with age (Abstract). Gerontologist *19:* 87 (1979c).

Harrison, D.E.: Competitive repopulation; a new assay for long-term stem cell functional capacity. Blood *55:* 77–81 (1980).

Harrison, D.E.: Measuring the functional abilities of stem cell lines; in Adler, Nordin, Immunological techniques applied to aging research, pp. 37–50 (UniScience, CRC Press, Florida 1981).

Harrison, D.E.: Experience with developing assays of physiological age; in Reff, Schneider, Biological markers of aging, pp. 2-12. NIH Publication No. 82-2221 (National Institutes of Health 1982).

Harrison, D.E.: Long-term erythropoietic repopulating abilities of stem cells from old, young, and fetal donors. J. exp. Med. *157:* 1496–1504 (1983).

Harrison, D.E.: Cell tissue transplantation: a means of studying the aging process; in Finch, Schreider, Handbook of the biology of aging, Chapter 14 (Van Nostrand & Reinhold, New York, in press, 1984).

Harrison, D.E.; Archer, J.R.; Astle, C.M.: The effect of hypophysectomy on thymic aging in mice. J. Immun. *129:* 2673–2677 (1982).

Harrison, D.E.; Astle, C.M.: Loss of stem cell repopulating ability with transplantation: effects of donor age, cell number and transplant procedure. J. exp. Med. *156:* 1767–1779 (1982).

Harrison, D.E.; Astle, C.M.; DeLaittre, J.A.: Loss of proliferative capacity in immunohemopoietic stem cells caused by serial transplantation rather than aging. J. exp. Med. *147:* 1526–1531 (1978).

Harrison, D.E.; Astle, C.M.; Doubleday, J.W.: Stem cell lines from old immunodeficient donors give normal responses in young recipients. J. Immun. *118:* 1223–1227 (1977).

Harrison, D.E.; Doubleday, J.W.: Normal functions of immunological stem cells from aged mice. J. Immun. *114:* 1314–1317 (1975).

Hirokawa, K.; Albright, J.W.; Makinodan, T.: Restoration of impaired immune function in aging animals. Clin. Immunol. Immunopathol. *5:* 371–376 (1976).

Hirokawa, K.; Sato, K.; Makinodan, T.: Restoration of impaired immune functions in aging animals. V. Long-term immunopotentiating effects of combined young marrow and newborn thymus grafts. Clin. Immunol. Immunopathol. *22:* 297–304 (1982).

Jaroslaw, B.N.; Suhrbier, K.M.; Fry, J.M.; Tyler, S.A.: In vitro suppression of immunocompetent cells by lymphomas from aging mice. J. natn. Cancer Inst. *54:* 1427–1432 (1975).

Kishimoto, S.; Shigemoto, S.; Yamamura, Y.: Immune response in aged mice: change of cell-mediated immunity with aging. Transplantation *15:* 455–459 (1973).

Kishimoto, S.; Takahama, T.; Mizumachi, H.: In vitro immune responses to the 2,4,6-trinitrophenyl determinant in aged C57BL/6J mice. J. Immun. *116:* 294–300 (1976).

Micklem, H.S.; Ford, C.E.; Evans, E.P.; Ogden, D.A.; Papworth, D.S.: Competition in vivo proliferation of foetal and adult haematopoietic cells in lethally irradiated mice. J. cell. Physiol. *79:* 293–298 (1972).

Myers, D.D.: Review of disease patterns and lifespan in aging mice: genetic and environmental interactions; in Bergsman, Harrison, Genetic effects on aging, pp. 41–54 (Liss, New York 1978).

Ogden, D.A.; Micklem, H.S.: The fate of serially transplanted bone marrow cell populations from young and old donors. Transplantation *22:* 287–293 (1976).

Price, G.B.; Makinodan, T.: Immunologic deficiencies in senescence. II. Characterization of extrinsic deficiencies. J. Immun. *108:* 413–417 (1972).

Ross, E.A.M.; Anderson, N.; Micklem, H.S.: Serial depletion and regeneration of the murine hematopoietic system. Implications for hematopoietic organization and the study of cellular aging. J. exp. Med. *155:* 432–444 (1982).

Russell, E.S.; Bernstein, S.E.: Proof of whole-cell implant in therapy of W-series anemia. Archs Biochem. Biophys. *125:* 594–597 (1968).

Szewczuk, M.R.; Dekruyff, E.A.; Weksler, M.E.; Siskind, G.W.: Ontogeny of B lymphocyte function. VIII. Failure of thymus cells from aged donors to induce the functional maturation of B lymphocytes from immature donors. Eur. J. Immunol. *10:* 918–923 (1980).

Tyan, M.L.: Age-related decrease in mouse T-cell progenitors. J. Immun. *118:* 846–851 (1977).

Tyan, M.L.: Marrow colony-forming units: age-related changes in responses to anti-θ-sensitive helper/suppressor stimuli. Proc. Soc. exp. Biol. Med. *165:* 354–360 (1980).

Tyan, M.L.: Old mice: marrow response to endotoxin or bleeding. Proc. Soc. exp. Biol. Med. *169:* 295–300 (1982a).

Tyan, M.L.: Effect of age on the intrinsic regulation of murine hemopoiesis. Mech. Age. Dev. *19:* 15–20 (1982b).

David E. Harrison, MD, The Jackson Laboratory, Bar Harbor, ME 04609 (USA)

Cellular Ageing: Selective Vulnerability of Cholinergic Neurones in Human Brain

David M. Bowen

Miriam Marks Department of Neurochemistry, Institute of Neurology, University of London, UK

Introduction

Evolution has led to the perfection of structure and function of the central nervous system and with it an increased life span. Thus, the non-dividing human nerve cells are exposed to ageing processes over an extended period. From about 50 years of age, there is a steady decrease in brain weight; and in some regions of the central nervous system, cell populations decrease. Microscopic examination reveals a few senile or neuritic plaques and a small proportion of neurones containing neurofibrillary degeneration or tangles, even in mentally normal individuals. The intensity of these histological changes is much greater over the age of 80 years. It seems that there are threshold values for the intensities of plaque and tangle formation before intellectual and personality deterioration occurs. Indeed, there is a correlation between intensity of plaque formation and the severity of mental incapacity [1, 2]. The ageing process seems to affect some individuals more than others. Thus, at one extreme, brain atrophy and histological changes are intensified, so that mental function is seriously affected. There is an acquired global defect in intellect which is clinically described as dementia. While the cause of this condition may be pathological (e.g. multi-infarct dementia of vascular origin), in the majority of cases the aetiology is unknown and may be ascribed to Alzheimer's disease, for the histological and clinical features were first described by the German physician of that name. It is one of the most devastating and commonest scourges of a long life [1–4].

During the last decade there has been considerable debate as to the importance and extent of neuronal loss in ageing. As I will show, biochemical studies in my laboratory have made a contribution to the problem and give an indication of the types of neurones affected as a result of ageing. Wherever possible I interpret changes in biochemical constituents in terms of their supposed cellular or subcellular identity and probable physiological action, although it is realised that this is an oversimplification, particularly for nerve cells and excitatory amino acid neurotransmitters.

Biochemistry of Normal Ageing Brain

Human Brain
Weight loss, atrophy and the increase in ventricular volume of the normal ageing human brain have been ascribed to nerve cell shrinkage or death. In order to test this concept we have assessed cellular and sub-cellular changes in autopsy brains of age range 50–100 years [5]. Those studied all appeared free of gross neurological or psychiatric disturbance. Since the temporal lobe has a role in memory we first examined this part of the brain. In order to avoid problems due to shrinkage, the entire lobe was analysed for the content of various representative biochemical constituents. The results suggested that depletion of nerve terminals occurs in ageing. These changes did not appear to be associated with a major structural disintegration, as wet weight, total protein, DNA, RNA and ganglioside (index of nerve cell membrane) content showed no significant age-related loss [5]. More recently, we have extended this earlier work to a larger group of samples and to an examination of different transmitter components of presynaptic and postsynaptic structures (fig. 1, 2). As well as the entire temporal lobe, normal neocortex from the temporal lobe and entire caudate nuclei have been analysed. The caudate nucleus was studied because of the involvement of the basal ganglia in dementing disease (Huntington's chorea and parkinsonism). In order to avoid the possibility of post-mortem artefacts, wherever possible, fresh normal appearing nervous tissue, obtained from surgical procedures, has been studied. The potential capacity of nerve endings to synthesise neuroactive transmitter has been measured by studying the appropriate synthetic enzymes and by direct measurement (fig. 1, footnote). Evoked release techniques have been used to measure physiologically active transmitter. Due to the post-mortem instability of the enzyme marker of catecholamine nerve endings (tyrosine hydroxylase), the activity in rat striatum has been used as a model for examining the effect of age on a representative component of dopamine nerve endings [9].

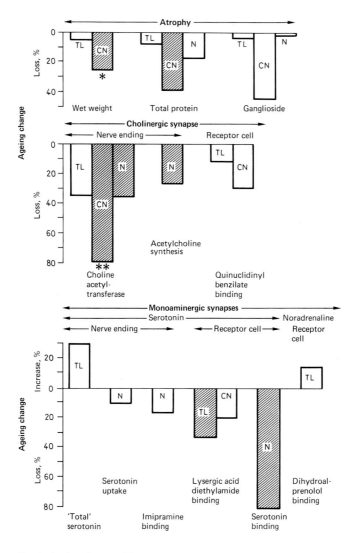

Fig. 1. Ageing changes (50–100 years, exceptions are below) in normal neocortex and the entire temporal lobe and caudate nucleus [from refs. 5–9]. 'Total' serotonin is the sum of the contents of serotonin and 5-hydroxyindole acetate. Ageing changes were determined by linear regression analysis (n was usually at least 16). The samples of neocortex were mostly from the temporal lobe. Acetylcholine synthesis (with U-^{14}C-glucose as precursor, which provides a genuine measure of the mass of transmitter synthesized [28]), uptake, evoked release (fig. 2) and choline acetyltransferase activity of neocortex were determined on samples obtained at neurosurgery (age range of 15–68 years). Synthesis was determined because the transferase probably does not control the rate of acetylcholine formation. For assaying synthesis the tissue

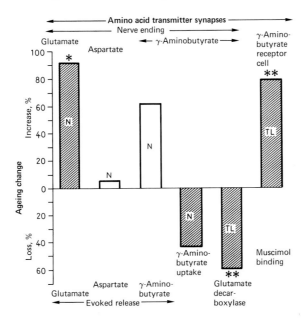

Fig. 2. Ageing changes (50–100 years, exceptions are in footnote to figure 1) in amino acid transmitter synapses [from refs. 5, 9, 11]. Details in legend to fig. 1.

In our limited study (fig. 1, 2) neither the wet weights of whole brain (data not shown) nor temporal lobe altered with increasing age. A slight loss of weight was apparent in the caudate nucleus. Protein and ganglioside content also appeared to decline but only in the caudate nucleus. In all tissues examined there was some evidence, especially in the caudate nucleus, that cholinergic terminals declined with ageing, without significant change in cholinoreceptive cells. The serotonin receptor cells altered with ageing only in the entire temporal lobe and neocortex. The serotonin nerve endings in neocortex, as well as the noradrenalin receptor cells in the entire temporal lobe, showed no age change. γ-Aminobutyrate synapses altered with ageing. The decrease found for uptake of γ-aminobutyrate by preparations of neo-

---

was mechanical by chopped to yield tissue prisms, in which the predominant intact structures are nerve endings [10]. Similar preparations were also used for measuring uptake and release of transmitters. TL = Entire temporal lobe; CN = entire caudate nucleus; N = neocortex. ▨ = Significant change: unmarked $p < 0.02$; *$p < 0.05$, and **$p < 0.01$; □ = non-significant trend.

Table I. Alzheimer's disease: biochemical measurements on two entire brain areas [from ref. 8, and unpublished results]

Potential indices of	Changes (content as % normal controls[1])	
	temporal lobe	caudate nucleus[2]
Cholinergic nerve endings (choline acetyltransferase)	35*	101
Glycolysis, initial reactions (aldolase, phosphohexoisomerase)	55–56*	n.d.
Number and shrinkage of nerve cells (ganglioside, enzyme associated with myelinated axons) Serotonin synapses (metabolite, metabolite/serotonin, lysergic acid diethylamide binding) Glycolysis, intermediate reaction (phosphoglycerate mutase)	60–70*	61–78
Glycolysis, intermediate reaction (triosephosphate isomerase) Mitochondria, microtubules (Mg^{2+} ATPase)	73–74*	n.d.
Atrophy (wet weight, total protein) Glycolysis, intermediate reaction (glyceraldehyde-3-phosphate dehydrogenase) Glycolysis, terminal reactions (pyruvate kinase; lactate dehydrogenase) Cell membranes (total ATPase)	77–81*	85–88

n.d. = not determined; myelinated axon enzyme is adenosine 2,3-cyclic nucleotide 3-phosphohydrolase and metabolite is 5-hydroxyindole acetate.
Bowen et al. [12] show evidence of loss of mitochondria (succinate dehydrogenase, 62% of control), DNA (74%), RNA (66%) and nerve endings (acetylcholine esterase, 48%).
*Identifies significant changes $p < 0.01$ (except $p < 0.02$ for total ATPase and pyruvate kinase); n = 16 (controls) and 17 (Alzheimer's disease).
[1] Age matched for variables that are agedependent (mean age of Alzheimer cases was 81, range 71–89).
[2] See [9]; not assayed for serotonin neurones, glycolytic enzymes or ATPases.

Table II. Alzheimer's disease: biochemical constituents in the entire temporal lobe showing no more than slight evidence of change [from ref. 8]

Markers in Alzheimer lobe showing:	Potential indices of
1 Abnormal age relationship	peptide modulation (angiotension-converting enzyme); synaptosomes (cyclic AMP-independent protein kinase); serotonin neurones (serotonin content); γ-aminobutyrate receptor cells (muscimol binding); glycolysis initial reactions (soluble hexokinase) and pentosephosphate cycle (glucose-6-phosphate dehydrogenase).
2 No difference from control	cholinoreceptive cells (atropine, quinuclidinyl benzilate binding); noradrenaline receptor cells (dihydroalprenolol binding); opiate receptor cells (naloxone binding); glycolysis, intermediate reactions (glycerate phosphokinase, enolase); pentose-phosphate cycle (6-phosphogluconate dehydrogenase); microsomes, synaptosomes (Na^+, K^+, ATPase); microsomes, synaptosomes, cytosol (total protein kinase), microsomes and myelinated axons (galactolipid).

cortex was smaller than for glutamate decarboxylase activity in the entire temporal lobe of older patients, suggesting that loss of γ-aminobutyrate terminals is less in younger patients. The pathophysiological effect of these changes may be reduced by an increase in the sensitivity of the receptor cell and in the concentration of physiologically active transmitter (see fig. 2). The concentration of physiologically active glutamate increased with ageing whereas that of aspartate was unaltered. Some 30 other constituents (table I, II) of the entire temporal lobe showed no significant age change (data not shown, see [5, 8]).

Thus, this work indicates that the cholinergic system in the caudate nucleus may be particular susceptible to ageing and that serotonin receptor cells are altered in the neocortex.

Rat and Human Brain: Vulnerable Cholinergic Interneurones

Choline acetyltransferase activity and acetylcholine synthesis in rat striatum declined with ageing whereas serotonin uptake by rat neocortex showed no significant age alteration [9, 13]. Tyrosine hydroxylase activity of rat striatum, has been determined for the first time in an ageing study in the presence of optimal concentrations of 'activating' substances. In agreement with other recent findings [14] there was no significant alteration with ageing [9]. Thus these data lay emphasis on the vulnerability in ageing brain of cholinergic nerve endings. There appears to be a decline in presynaptic cholinergic function with age, with no alteration or even a slight decline [15] in cholinoreceptive cells. Hence cholinergic transmission probably declines with ageing. The loss of choline acetyltransferase activity seemed to be less in the temporal cortex and temporal lobe than in the caudate nucleus. Moreover, little or no age-change occurs in frontal cortex [4, 8, 9; see also 16]. The larger loss of transferase activity from the striatum is in agreement with the results of our study of the effect of age on regional acetylcholine synthesis in rat brain [13]. The marked age-related losses of presynaptic cholinergic activity from the neostriatum in both species suggests that interneurones releasing acetylcholine are especially vulnerable in the ageing brain [for review see 9].

Preservation of Some Synaptic Structures

By contrast with the acetylcholine and γ-aminobutyrate systems, there is evidence of a marked age-related decline in the serotonin receptor cells in cortex [17] but not of significant loss of serotonin nerve endings in either human or rat brain. The preservation of these nerve terminals in neocortex of both species suggests that the ascending projections to the neocortex from serotonergic perikarya in the brain stem [18] are preserved in the normal ageing brain. Similarly, there is no biochemical evidence of an age-related decline in noradrenergic activity in cortex or locus coeruleus, although reduced cell counts have been reported with ageing in the latter region [for a review see 9]. The ascending nigrostriatal dopamine pathway in rat brain seems unaffected because the activities of neither tyrosine hydroxylase nor dopamine decarboxylase [19] alter with age.

The relative stability with ageing of most markers of nerve terminals in the neocortex suggest that there may be no major degenerative changes in certain specific groups of neurones with long axons that project to the neocortex. These include the large cholinergic perikarya in the nucleus basalis of Meynert [20, 21] and serotonin and noradrenalin perikarya in the brain stem. γ-Aminobutyrate, glutamate, aspartate and dopamine transmission may also be relatively spared from age-related loss. Further work is needed, however, on other biochemical markers and cell counts, before the lack of such changes can be ruled out.

Biochemistry of Alzheimer Brain

Two types of Alzheimer's disease are arbitrarily classified on the basis of age (before or after 65 years). Since both show similar neuropathological changes [3] in this review both are referred to as Alzheimer's disease. In Alzheimer's disease neuronal fall-out occurs from the hippocampus but there is contradictory evidence about the extent of cell loss from the neocortex. Moreover, although significant atrophy and loss of brain hemispheric volume occurs there is considerable variation in the severity of cerebral atrophy. The interpretation of such changes is complicated for some cortical neurones are growing, while others are degenerating [for reviews see 2, 3, 7, 12, 22]. This emphasises the importance of showing the overall extent to which biochemical indices of nerve cells and atrophy are affected.

Cellular and Sub-Cellular Changes

Biochemical estimation of changes were first made in the frontal neocortex and a region devoid of senile histopathology (caudate nucleus). This work commenced in 1972, utilizing brains from Runwell Psychiatric Hospital, kindly provided by Prof. *Corsellis* (the first systematic examination of the incidence of plaque and tangle formation in a 'population' is described by *Corsellis* [1], who studied patients that died in Runwell Hospital). Cases studied were histologically proven examples of the senile form of Alzheimer's disease and age-matched controls. In 1975 we first showed marked reduction in cortical choline acetyltransferase activity [7]. The depletion in activity for individual cases was related to the intensity of plaque and tangle formation [23, 24]. The enzyme was not significantly reduced in the caudate nucleus compared to age-matched controls. The cholinergic defect in autopsy and biopsy samples is now well documented [for reviews see 3, 20]. In patients it significantly correlates with the degree of dementia determined shortly before diagnostic craniotomy [*Neary* et al., unpublished].

To make allowance for brain shrinkage, the content of various biochemical constituents have been determined in the entire temporal lobe and caudate nucleus. Table I shows that the caudate nucleus in the senile form of Alzheimer's disease was not significantly affected in comparison to age-matched controls. However, a 20% shrinkage of the temporal lobe was associated with a loss of about one-third of the nerve cell component, ganglioside, and the enzyme associated with myelinated axons. There was a 52% loss of acetylcholine esterase. The 65% loss of choline acetyltransferase activity was the largest of all changes found in the 40 constituents measured (table I, II) [12].

Reduced activities of two other enzymes of nerve endings (glutamate and aromatic amino acid decarboxylase) are thought to be related to terminal agonal state [7, 25, 26]. Other biochemical findings suggested that in addition to loss of nerve cell population there was shrinkage of neurones and of cell processes. Only slight or no change occurred in indices of capillaries and glial cells [7, 8, 12].

In addition to cholinergic loss, there are other more subtle selective biochemical changes in the Alzheimer temporal lobe in comparison to age- and sex-matched controls. Abnormal constituents include the serotonin content and activity of soluble hexokinase (table II). The latter change is notable for the activities of two other glycolytic enzymes (aldolase and phosphohexoisomerase) are about halved in Alzheimer temporal lobe (table I). No more than slight changes were detected in the activities of most other enzymes involved in glucose and energy metabolism and protein phosphorylation. Overall glucose metabolism may be reduced as a consequence of altered mitochondria, assessed by succinate dehydrogenase activity of unfractionated homogenates of the temporal lobe (table I, footnote). The content of the serotonin metabolite, 5-hydroxyindole acetate, is reduced by some 33%. Although this is consistent with the finding of an abnormal serotonin content, it is unclear whether or not these other changes (i.e. in serotonin metabolism and glycolytic enzymes) are critical factors in Alzheimer's disease, for the markers may be affected by changes caused by terminal hypoxia [8]. Another marker of the serotonin synapse (binding of lysergic acid diethylamide by membranes from the temporal lobe) was reduced by 33% (table I). There was no significant change in the binding of non-serotonin-like ligands (table II).

Vulnerable Cholinergic Neurones with Long Axons

Investigation of patients with presenile dementia sometimes includes brain sampling by diagnostic craniotomy [5]. In a study of such samples from the frontal lobe, 4 were from patients who were probably biopsied within a year of onset of the disease [4]. These showed clear histological evidence of Alzheimer's disease and the mean choline acetyltransferase activity was significantly reduced from control by 61%. Since Alzheimer's disease usually ends in death 5 years after onset, this finding supports our view [8] that the cholinergic defect is an early change. In biopsy samples predominantly from the temporal lobe, synthesis of acetylcholine was directly measured [10, 27, 28]. It was markedly reduced only in the demented patients who exhibited the histological features of Alzheimer's disease. Similar reductions were observed for measurements made in temporal and frontal lobe and in the presence of either a low (5 mM) or high (31 mM) potassium concentration. Thus, the alteration in synthesis did not seem to be related to either the brain region or to the metabolic activity of the tissue. Choline uptake and choline acetyltransferase activity were also significantly reduced, to approximately the same extent as acetylcholine synthesis (table III).

Although changes in choline acetyltransferase activity in Alzheimer's disease have frequently been interpreted as indicating changes in acetylcholine synthesis, our work provides the first direct evidence of a reduced capacity for synthesis in this disease [10, 27, 28]. In animal studies, choline uptake and acetylcholine synthesis generally relate closely, although there is not obligatory coupling. The activity of the transferase may change independently. Thus, the similar decreases in these parameters is best interpreted as reflecting the loss of functionally-active cholinergic nerve endings with associated reductions in the three markers which are closely related to these structures. An alternative explanation could arise if choline acetyltransferase becomes rate limiting in the diseased human tissue. However, under such circumstances it is likely that synthesis would be more affected at the higher rate (i.e. in 31 mM potassium) whereas the proportional response to potassium was found to be largely preserved in the disease tissue. Several mechanisms could produce loss of function of cholinergic cells. A strong possibility being the death of whole cholinergic cells whose cell bodies are in the nucleus basalis (table IV) [20].

Selective Vulnerability of Cholinergic Neurones

Table III shows that loss of cholinergic activity from the neocortex in Alzheimer's disease is accompanied by significant but usually smaller reductions in markers of serotonin nerve endings in both autopsy and biopsy samples. Serotonin receptor cells were also significantly reduced but in only severely affected cases [22]. There was no loss of glutamate decarboxylase activity. Moreover, releasable γ-aminobutyrate, glutamate and aspartate was not significantly different from control values (table III). The concentration of the dopamine metabolite, homovanillate, in the cerebrospinal fluid of biopsy-confirmed Alzheimer patients was not reduced (table III). Thus, amino acid nerve endings in the neocortex and dopamine terminals in the basal ganglia do not seem to be affected. Elsewhere, indices of noradrenalin neurones and seven neuropeptides have been assayed in Alzheimer and control brain tissue. Of these only the noradrenergic markers and somatostatin concentration show abnormalities and even these are not generally as exten-

Table III. Acetylcholine, serotonin, dopamine and amino acid transmitter-containing nerve endings

Nerve ending	Measurement	Specimens assayed	Alzheimer samples	Control samples
Acetylcholine	choline acetyltransferase	biopsy cortex (pmol/mg/min)	34 ± 19 (15)**	83 ± 24 (13)
		autopsy cortex (pmol/mg/min)	40 ± 17 (15)**	83 ± 24 (12)
		entire temporal lobe (nmol/lobe/min)	173 ± 79 (8)**	442 ± 170 (6)
	choline uptake	biopsy cortex (pmol/mg/min)	1.72 ± 0.58 (6)**	3.04 ± 0.33 (6)
	acetylcholine synthesis	biopsy cortex (dpm/mg/min)	3.4 ± 1.2 (13)**	7.3 ± 1.4 (20)
Serotonin	serotonin uptake	biopsy cortex (fmol/mg/min)	298 ± 100 (6)**	1065 ± 360 (5)
	imipramine binding	autopsy cortex (fmol/mg)	48.5 ± 9.0 (13)*	59.1 ± 9.0 (7)
	total serotonin	entire temporal lobe (μg/lobe)	45 ± 9 (5)*	55 ± 11 (15)
Dopamine	homovanillate	lumbar cerebrospinal fluid (ng/ml)	29 ± 12 (5)	33 ± 16 (12)
Glutamate	evoked transmitter release	biopsy cortex (nmol/mg)	9.87 ± 2.55 (7)	7.02 ± 2.33 (6)
Aspartate	evoked transmitter release	biopsy cortex (nmol/mg)	2.16 ± 0.55 (7)	1.85 ± 0.55 (6)
Aminobutyrate	evoked transmitter release	biopsy cortex (nmol/mg)	1.69 ± 0.24 (7)	1.81 ± 0.57 (6)
	glutamate decarboxylase	biopsy cortex (pmol/mg/min)	1298 ± 205 (4)	1333 ± 467 (8)

From Bowen and colleagues [4, 22, 28–31].
Total serotonin, see figure 1. Number of specimens shown in parentheses.
Significant differences from the mean control value; $^*p < 0.05$, $^{**}p <$ at least 0.01.

Most of the samples are from the temporal lobe. Biopsy cortex and lumbar cerebrospinal fluid are from Alzheimer patients of presenile age, temporal lobes are from late-onset Alzheimer cases (died age 80 and over) and autopsy cortex is from Alzheimer cases of age 51–93 (mean age 69 ± 14). Control samples are matched to dementia specimens with respect to sampling delay after death and age, where appropriate. The data is expressed per mg protein.

Table IV. Cell counts in the nucleus basalis of Meynert and cholinergic activity

Diagnosis	Mean count per field	Choline acetyltransferase activity, (pmol/min/mg protein)		
		frontal cortex	temporal cortex	nucleus basalis
Control	49	77	91	1363
Alzheimer's disease	25**	34*	28**	432**

From *Wilcock* et al. [21], except the enzyme activity of subcortex [32].
Significant differences from control: *p < 0.02, **p < 0.001 for n = 5–19 (control) and 6–16 (Alzheimer's disease).

sive and consistently observed as the cholinergic deficit [for refs. see 32, 33]. These observations, together with the relationship between neurohistopathology, mental state and the cholinergic indices suggest that the cholinergic deficit is an important change of clinical relevance. No similar relationship has yet been established for any other neurotransmitter system in Alzheimer's disease.

The technique we developed for measuring acetylcholine synthesis has allowed direct estimation of overall glucose oxidation, by trapping radioactive CO_2 produced by tissue prisms incubated with labelled glucose. The production of radioactive CO_2 by Alzheimer biopsy samples is significantly higher compared with controls [10] but specific labelling of the total free amino acids in the diseased tissue is unaltered. Hence, it is likely that the increase in production of labelled CO_2 can be attributable to an unexpected increase in glucose utilization. Possibly only in nerve endings, because these are the predominant intact structure of tissue prisms [10]. The content of energy-rich nucleotides in tissue prisms from temporal cortex has been determined by fluorescence analysis of reduced nicotinamide adenine dinucleotide. In Alzheimer compared to control samples similar reductions were found in the content (per mg prism protein) of ATP, ADP and AMP (reduced by 21, 21 and 28%, respectively). The overall energy balance of the tissue prisms, as assessed by the adenylate energy charge, was unchanged from the control value. One explanation for these new observations is that they reflect recovery in vitro of the tissue from a previously undescribed pathological state in situ [*Sims* et al., unpublished].

Conclusions

Cholinergic, serotonin and noradrenergic innervation probably constitute less than 12% of the total nerve terminals in the human neocortex [for reviews see 10, 22]. Hence the large reduction in the entire Alzheimer temporal lobe of nerve cell components common to all nerve cells (ganglioside and the enzyme associated with myelinated axons as well as adenine nucleotides of cortex) suggests that there is a major loss of so far unidentified nerve cells or their processes from the neocortex. From the present results, it can be speculated that the affected cells possess serotonin receptors and relatively high activities of aldolase and phosphohexoisomerase. Cell loss from the neocortex is consistent with some morphological observations and reduced somatostatin-immunoreactivity, associated with intrinsic cortical neurones [for review see 22]. There is already evidence that the loss is selective because indices of glutamate, aspartate, γ-aminobutyrate (table III) and most neuropeptide-containing [32] intrinsic neurones are not significantly reduced. As major descending projections from the neocortex (e.g. glutamate neurones projecting to the thalamus and neostriatum and amygdala of the limbic system) have yet to be examined, the physiological implication of Goth tangle-bearing neurones cell loss from the neocortex in Alzheimer's disease remains unclear [22].

The present data demonstrate that several measures of presynaptic cholinergic nerve terminals are markedly reduced in both the neostriatum from aged humans and rats and cortical brain tissue from Alzheimer patients. This indicates that in later life, cholinergic interneurones are affected, as well as cholinergic neurones with long axons and cell bodies in the basal forebrain. Many other substances and processes have been assessed, including six other classical transmitters, but none of these show uniform and marked losses. Thus, it is suggested that studies of cellular ageing should focus on cholinergic neurones. Possible properties of cholinergic neurones that may render them vulnerable in ageing brain have already been identified. These include a functional activity tightly linked to carbohydrate utilisation [34], sensitivity towards growth promoting or trophic factors [35] and the possibility that certain cholinergic dendrites form senile plaques [20].

Since releasable glutamate seems to be present in excess in the neocortex of ageing human brain (fig. 2) the substance may have an 'excitotoxic' action, triggering neuronal degeneration [36]. Similar studies on affected regions (table IV) not available from surgical procedures may now be possible [11, 36, 37].

Therapeutic Intervention

The belief that dementia was due to vascular insufficiency led, in the past, to the use of drugs which might improve the presumed cerebral ischaemia. Vasodilators, anticoagulants and hyperbaric oxygen were used for this purpose, but with no consistent benefit [38]. In 1977 we first reported that postsynaptic muscarinic receptors are intact in Alzheimer brain [15]. This has been confirmed, also for the nicotinic receptor [39]. Thus, there has been an attempt to increase endogenous acetylcholine synthesis within the brain of patients by feeding choline or lecithin. This strategy is based on the finding that the concentration of the transmitter in the brain has been found to be increased by raised dietary choline intake [for reviews see 38, 40]. In vitro studies on human tissue in this laboratory [40] indicate that an exceptionally high extracellular concentration of choline is required for the remaining cortical nerve endings in Alzheimer brain to synthesise a normal amount of acetylcholine. This makes it most unlikely that dietary choline would reverse the deficit [28]. Blockade of the muscarinic autoreceptor on cholinergic nerve terminals provides another approach for increasing transmitter synthesis [3]. However, although we found that acetylcholine synthesis in rat brain can be increased by such drugs by nearly 50% [40, 41] the increment in human samples is only 12% [40]. Moreover, our study of various antimuscarinic agents has laid emphasis on the pharmacological similarity between the pre- and postsynaptic muscarinic receptor [41]. Another factor to be considered is the control of choline entry into the nerve endings. In experimental animals the capacity of the high-affinity choline uptake system is linked to neuronal activity. For example, anaesthetics and drugs decreasing acetylcholine synthesis reduce the uptake of choline. Thus, a defect in neuronal energy metabolism [10] or a block of axonal flow from the cell body [38] could be the key factor affecting the cholinergic terminals. Treatment with choline agonists or centrally acting anticholinesterase may therefore be only partially effective in demented patients, but the possibility of combining dietary choline with a drug stimulating neuronal metabolism [42] is yet to be tested.

In parkinsonism, reduction in the dopaminergic cell population of the substantia nigra leads to loss of dopaminergic terminals in the striatum. In Stockholm an attempt has been made to correct for this deficiency by transplanting a patient's own adrenal tissue into the caudate nucleus [for review see 43]. Regeneration of central cholinergic connections has been observed in experimental animals with intracerebral neural transplants [44]. Thus, there is hope that in the future it may be possible to replace the cholinergic neurones that are thought to be lost in Alzheimer's disease. There is an evidence of glutamate receptor supersensitivity in Alzheimer's disease [45]. Neurones with glutamate receptors of the N-methyl-D-aspartate subtype include cholinergic cells [46]. Quinolinic acid (2,3-pyridine dicarboxylic acid), an endogenous neuroexcitotoxin, acts selectively at this receptor [47] in the nucleus basalis [48]. I speculate that a substance which prevents quinolinate toxicity might retard cholinergic cell loss in Alzheimer patients.

Acknowledgments

The work carried out in this laboratory would have not been possible without the generous collaboration, help and support of a large number of people, to whom the author is very grateful. Financial support was from the Medical Research Council, Brain Research Trust, Miriam Marks Charitable Trust, Wellcome Foundation and Sandoz Pharmaceuticals.

References

1   Corsellis, J.A.N.: Mental illness and the ageing brain (Oxford University Press, Oxford 1962).
2   Tomlinson, B.E.: The structural and quantitative aspects of the dementias; in Roberts, Biochemistry of dementia, pp. 15–52 (Wiley, Chichester 1980).
3   Bowen, D.M.: Alzheimer disease; in Thompson, Davison, Molecular basis of neuropathology, pp. 649–665 (Edward Arnold, London 1981).
4   Bowen, D.M.; Benton, J.S.; Spillane, J.A.; Smith, C.C.T.; Allen, S.J.: Choline acetyltransferase activity and histopathology of frontal neocortex from biopsies of demented patients. J. neurol. Sci. *57:* 191–202 (1982).
5   Bowen, D.M.; Smith, C.B.; White, P.; Goodhardt, M.J.; Spillane, J.A.; Flack, R.H.A.; Davison, A.N.: Chemical pathology of the organic dementias. I. Validity of post-mortem measurements on human post-mortem specimens. Brain *100:* 397–426 (1977).
6   Bowen, D.M.; Smith, C.B.; Davison, A.N.: Molecular changes in senile dementia. Brain *96:* 849–856 (1973).
7   Bowen, D.M.; Smith, C.B.; White, P.; Davison, A.N.: Neurotransmitter-related enzymes and indices of hypoxia in senile dementia and other abiotrophies. Brain *99:* 459–496 (1976).
8   Bowen, D.M.; White, P.; Spillane, J.A.; Goodhardt, M.J.; Curzon, G.; Iwangoff, P.; Meier-Ruge, W.; Davison, A.N.: Accelerated ageing or selective neuronal loss as an important cause of dementia? Lancet *i:* 11–14 (1979).
9   Allen, S.J.; Benton, J.S.; Goodhardt, M.J.; Haan, E.A.; Sims, N.R.; Smith, C.C.T.; Spillane, J.A.; Bowen, D.M.; Davison, A.N.: Biochemical evidence of selective nerve cell changes in the normal ageing human and rat brain. J. Neurochem. *41:* 256–265 (1983).
10 Sims, N.R.; Bowen, D.M.; Davison, A.N.: ^{14}C acetylcholine synthesis and ^{14}C carbon dioxide production from U-^{14}C glucose by tissue prisms from human neocortex. Biochem. J. *196:* 867–876 (1981).

11 Smith, C.C.T.; Bowen, D.M.; Davison, A.N.: The evoked release of endogenous amino acids from tissue prisms of human neocortex. Brain Res. *269:* 103–109 (1983).

12 Bowen, D.M.; Smith, C.B.; White, P.; Flack, R.H.A.; Carrasco, L.H.; Gedye, J.L.; Davison, A.N.: Chemical pathology of the organic dementias. II. Quantitative estimation of cellular changes in post-mortem brains. Brain *100:* 427–453 (1977).

13 Sims, N.R.; Marek, K.L.; Bowen, D.M.; Davison, A.N.: Production of ^{14}C-acetylcholine and ^{14}C-carbon dioxide from U ^{14}C-glucose in tissue prisms from ageing rat brain. J. Neurochem. *38:* 488–492 (1982).

14 Osterburg, H.H.; Donahue, H.G.; Severson, J.A.; Finch, C.E.: Catecholamine levels and turnover during ageing in brain regions of C57BL/6J mice. Brain Res. *224:* 337–352 (1981).

15 White, P.; Hiley, C.R.; Goodhardt, M.J.; Carrasco, L.H.; Keet, J.P.; Williams, I.E.I.; Bowen, D.M.: Neocortical cholinergic neurons in elderly people. Lancet *i:* 668–671 (1977).

16 Rossor, M.N.; Iversen, L.L.; Johnson, A.J.; Mountjoy, C.Q.; Roth, M.: Cholinergic deficit in frontal cerebral cortex in Alzheimer's disease is age-dependent. Lancet *ii:* 1422 (1981).

17 Shih, J.C.; Young, H.: The alteration of serotonin binding sites in aged human brain. Life Sci. *23:* 1441–1448 (1978).

18 Benton, J.S.; Bowen, D.M.; Allen, S.J.; Haan, E.A.; Davison, A.N.; Neary, D.; Murphy, R.P.; Snowden, J.S.: Alzheimer's disease as a disorder of isodendritic core. Lancet *i:* 456 (1982).

19 Reis, D.J.; Ross, R.A.; Joh, T.H.: Changes in the activity and amounts of enzymes synthesizing catecholamines and acetylcholine in brain, adrenal medulla and sympathetic ganglia of aged rat and mouse. Brain Res. *137:* 465–474 (1977).

20 Price, D.L.; Whitehouse, P.J.; Struble, R.G.; Clark, A.W.; Coyle, J.T.; DeLong, M.R.; Hedreen, J.C.: Basal forebrain cholinergic systems in Alzheimer's disease and related dementias. Neurosci. Commentaries *1:* 84–92 (1982).

21 Wilcock, G.K.; Esiri, M.M.; Bowen, D.M.; Smith, C.C.T.: The nucleus basalis in Alzheimer's disease. Cell counts and cortical biochemistry. Neuropath. appl. Neurobiol. *9:* 175–179 (1983).

22 Bowen, D.M.; Allen, S.J.; Benton, J.S.; Goodhardt, M.J.; Haan, E.A.; Palmer, A.M.; Sims, N.R.; Smith, C.C.T.; Spillane, J.A.; Esiri, M.M.; Neary, D.; Snowden, J.S.; Wilcock, G.K.; Davison, A.N.: Biochemical assessment of serotonergic and cholinergic dysfunction and cerebral atrophy in Alzheimer's disease. J. Neurochem. *41:* 266–272 (1983).

23 Perry, E.K.; Tomlinson, B.E.; Blessed, G.; Bergmann, K.; Gibson, P.H.; Perry, R.H.: Correlation of cholinergic abnormalities with senile plaques and mental test scores in senile dementia. Br. med. J. *ii:* 1457–1459 (1978).

24 Wilcock, G.K.; Esiri, M.M.; Bowen, D.M.; Smith, C.C.T.: Alzheimer's disease: correlation of cortical choline acetyltransferase activity with the severity of dementia and histological abnormalities. J. neurol. Sci. *57:* 407–417 (1982).

25 Bowen, D.M.; White, P.; Flack, R.H.A.; Smith, C.B.; Davison, A.N.: Brain-decarboxylase activities as indices of pathological change in senile dementia. Lancet *i:* 1247–1249 (1974).

26 Bowen, D.M.; Goodhardt, M.J.; Strong, A.J.; Smith, C.B.; White, P.; Branston, A.N.; Symon, L.; Davison, A.N.: Biochemical indices of brain structure, function and 'hypoxia' in cortex from baboons with middle cerebral artery occlusion. Brain Res. *117:* 503–507 (1976).

27 Sims, N.R.; Bowen, D.M.; Smith, C.C.T.; Flack, R.H.A.; Davison, A.N.; Snowden, J.S.;

Neary, D.: Glucose metabolism and acetylcholine synthesis in relation to neuronal activity in Alzheimer's disease. Lancet *i:* 333–336 (1980).

28 Sims, N.R.; Bowen, D.M.; Allen, S.J.; Smith, C.C.T.; Neary, D.; Thomas, D.J.; Davison, A.N.: Presynaptic cholinergic dysfunction in patients with dementia. J. Neurochem. *40:* 503–509 (1983).

29 Spillane, J.A.; White, P.; Goodhardt, M.J.; Flack, R.H.A.; Bowen, D.M.; Davison, A.N.: Selective vulnerability of neurones in organic dementia. Nature, Lond. *266:* 558–559 (1977).

30 Bowen, D.M.; Sims, N.R.; Benton, J.S.; Curzon, G.; Davison, A.N.; Neary, D.; Thomas, D.J.: Treatment of Alzheimer's disease: a cautionary note. New Engl. J. Med. *305:* 1016 (1981).

31 Smith, C.C.T.; Bowen, D.M.; Sims, N.R.; Neary, D.; Davison, A.N.: Amino acid release from biopsy samples of temporal neocortex from patients with Alzheimer's disease. Brain Res. *264:* 138–141 (1983).

32 Rossor, M.N.; Garrett, N.J.; Johnson, A.L.; Mountjoy, C.Q.; Roth, M.; Iversen, L.L.: A post-mortem study of the cholinergic and GABA systems in senile dementia. Brain *105:* 313–320 (1982).

33 Yates, C.M.; Harman, A.J.; Rosie, R.; Sheward, J.; Sanchez de Levy, G.; Simpson, J.; Maloney, A.F.J.; Gordon, A.; Fink, G.: Thyrotropin-releasing hormone, luteinizing hormone-releasing hormone and substance P immunoreactivity in post-mortem brain from cases of Alzheimer-type dementia and Down's syndrome. Brain Res. *258:* 45–52 (1983).

34 Blass, J.P.; Gibson, G.E.; Shimada, M.; Kihara, T.; Watanabe, M.; Kurinoto, K.: Brain carbohydrate metabolism and dementias; in Roberts, Biochemistry of dementia, pp. 121–134 (Wiley, Chichester 1980).

35 Appel, S.H.: A unifying hypothesis for the cause of amyotrophic lateral sclerosis, Parkinsonism and Alzheimer's disease. Ann. Neurol. *10:* 499–505 (1981).

36 Bowen, D.M.; Smith, C.C.T.; Davison, A.N.: Excitotoxicity in ageing and dementia; in Fuxe, Schwartz, Roberts, Excitotoxins, pp. 354–362 (Pergamon Press, Oxford 1983).

37 Bowen, D.M.; Sims, N.R.; Lee, K.A.P.; Marek, K.L.: Acetylcholine synthesis and glucose oxidation are preserved in human brain obtained shortly after death. Neurosci. Lett. *31:* 195–199 (1982).

38 Bowen, D.M.; Davison, A.N.: The failing brain. J. chron. Dis. *36:* 3–13 (1983).

39 Lang, W.; Henke, H.: Cholinergic receptor binding and autoradiography in brain of non-neurological and senile dementia of Alzheimer type patients. Brain Res. (in press).

40 Marek, K.L.; Bowen, D.M.; Sims, N.R.; Davison, A.N.: Stimulation of acetylcholine synthesis by blockade of presynaptic muscarinic inhibitory autoreceptors: observations in rat and human brain preparations and comparison with the effect of choline. Life Sci. *30:* 1517–1524 (1982).

41 Bowen, D.M.; Marek, K.L.: Evidence for the pharmacological similarities between the central presynaptic muscarinic autoreceptor and postsynaptic muscarinic receptors. Br. J. Pharmacol. *75:* 367–372 (1982).

42 Bowen, D.M.; Davison, A.N.: Neurotransmitter and neurophysiological changes in relation to pathology in senile dementia or Alzheimer's disease; in Stein, The psychobiology of aging: problems and perspectives, pp. 331–346 (Elsevier North-Holland, Amsterdam 1980).

43 Marx, J.L.: Transplants as guides to brain development. Science *217:* 340–342 (1982).

44 Bjorklund, A.: Regeneration of central monoaminergic and cholinergic connections as

revealed in experiments with intracerebral neural transplants; in Berry, 6th European Anatomical Congress, Hamburg 1981. Growth and regeneration of axons in the nervous system. Bibthca anat., vol. 23, pp. 93–94 (Karger, Basel 1982).

45 Bowen, D.M.; Davison, A.N.; Francis, P.T.; Palmer, A.M.; Pearce, B.R.: Neurotransmitter dysfunction in Alzheimer's dementia. Relationship to histopathological features; in Rose, Recent advances in dementias (in preparation).

46 Lehmann, J.; Scatton, B.: Characterization of the excitatory amino-acid receptor mediated release of [^{3}H]acetylcholine from rat striatal slices. Brain Res. *252:* 77–89 (1982).

47 Perkins, M.N.; Stone, T.W.: Pharmacology and regional variations of quinolinic acid-evoked excitations in the rat central nervous system. J. Pharmac. exp. Ther. *226:* 551–557 (1983).

48 Schwarcz, R.; Köhler, C.: Differential vulnerability of central neurones of the rat to quinolinic acid. Neurosci. Lett. *38:* 85–90 (1983).

D.M. Bowen, PhD, Miriam Marks Department of Neurochemistry, Institute of Neurology, University of London, Queen Square, London WC1N 3BG (England)

The Unsolved Problem of Cellular Ageing

Robin Holliday

National Institute for Medical Research, London, England

Introduction

Thirty years ago, *Medawar* [43] published a review on the ageing of animals under the title 'An unsolved problem in biology'. At that time the study of the ageing of cells or populations of cells in the laboratory had not been initiated, but in the ensuing 30 years, it has become possible to carry out experiments on cells in culture, particularly those derived from human tissue or chick embryos, as well as with several protozoa and fungi.

Following the early work on cultured chick cells by *Carrel* and his associates [for a review, see 75] and the establishment of permanent cell lines from human tumours, it was widely believed that all vertebrate cells capable of division could be propagated indefinitely in culture. This view was challenged by *Hayflick and Moorhead* [14], following the earlier demonstration by *Swim and Parker* [68] that there is a limit to the serial subculture of human connective tissue cells, generally referred to as fibroblasts. *Hayflick* [12] fully documented this phenomenon by carrying out a series of control experiments, which established for the first time that the finite growth of human diploid fibroblasts was intrinsic to the cells themselves and could not be readily explained by environmental factors, such as faulty medium, or the presence of infectious agents. He proposed that these intrinsic changes in fibroblasts are a manifestation of ageing at the cellular level, an interpretation which has been widely, although not universally, accepted. He also clarified some of the basic differences between heteroploid transformed cells, which grow indefinitely, and diploid cells which do not.

The results of *Carrel* with chick cells have never been substantiated [75], and many subsequent studies have demonstrated that chick embryo fibroblasts have an even shorter growth potential in culture than human cells. A remarkable feature of populations of human and chick fibroblasts is the apparent absence of any subpopulation of transformed cells, since if any of these arose during serial subculture, they would survive the senescence of the diploid cells and form permanent lines. Such permanent lines can occasionally be obtained from human diploid cell populations treated with SV40 virus [26], but they consistently emerge from untreated rodent or rabbit cell populations. Three major problems arise from these studies. First, what are the molecular mechanisms which limit the proliferation of cultures of fibroblasts? Second, what is the basis of the difference in growth potential of diploid and transformed cells? Third, what genetic or other factors determine the longevity of cells from different species and the likelihood of spontaneous transformation?

None of these problems has yet been solved. In this critical review I assess some of the theoretical and experimental advances which have so far been made, with particular reference to the studies with human fetal lung fibroblast strain MRC-5 in this laboratory in the last 10 years.

Validity of Hayflick's Experimental Model

Although the finite growth of human fibroblasts has been demonstrated on innumerable occasions in many laboratories, there is continuing controversy surrounding the interpretation that it is due to cellular ageing. *Martin* et al. [41] and *Bell* et al. [1] have proposed that the cessation of growth is due to a process of terminal differentiation in which the cells enter a non-cycling state. There are several reasons for believing this interpretation to be incorrect. Firstly, the fibroblast is already a differentiated connective tissue cell capable of collagen synthesis. In normal development, differentiation is indeed terminal and it is not observed that one type of differentiated cell changes into another. Second, examination of proteins by two-dimensional gels have not revealed other than slight changes in the pattern, although newly differentiated cells might be expected to synthesize significant amounts of one or more new 'luxury' proteins [58]. Third, there is no evidence that late passage cells adopt a uniform, distinct phenotype. Instead, the cells vary considerably in size, shape, number of nuclei and degree of granularity. Their structure deteriorates, and they frequently round up and detach from the surface. Fourth, premature ageing syndromes, or those with karyotype instability or defects in DNA repair, strikingly reduce the longevity of cultured fibroblasts [40, 72, 73]. However, these individuals show, for the most part, normal development and it is hard to see why their cells in culture should be strikingly altered in temporal differentiation. Finally, one might ask why there should be such a long period of normal proliferation (50–70 population doublings) before differentiation takes place. Cultured cells which are known to be predetermined to differentiate, such as epidermal keratinocytes, usually have only limited growth in culture [67].

Some cell biologists have maintained (often in open discussion, rather than in print) that the finite growth of diploid fibroblasts is merely an artifact of laboratory culture conditions, and that if an ideal medium could be found, the cells would grow indefinitely. However, this does not mean that the cells do not age under the conditions which have so far been used; rather it implies that the cells are in a state of 'unbalanced growth'. Unbalanced growth, or the gradual breakdown of the normal cellular homeostatic mechanisms, is itself a valid theory of ageing. As *Hayflick and Moorhead* [14] pointed out long ago, the limited growth could not be due to the simple dilution of an essential cellular constituent, but it might be due to a progressive imbalance in cellular constituents. It is possible, for instance, that the formation of certain organelles does not keep pace with cell division, so that they slowly become depleted [51]. Alternatively, they may be overproduced, until finally there is a lethal imbalance. In *Podospora*, there is evidence that cellular senescence is associated with the proliferation of mitochondria with abnormal DNA sequences [74], but so far, there is no evidence that the same changes occur in cultured vertebrate cells. However, there is some evidence that changes in certain nuclear DNA sequences occur during in vitro ageing [60]. This raises the possibility that ageing could be due to a failure of normal DNA maintenance mechanisms. These suppositions are now testable, using the new procedures of DNA technology. However, even if positive results were obtained, the problem of the origin of these changes and why they occur in some cells and not others would remain unresolved. In the same way, the discovery of the changes in mitochondrial DNA of *Podospora* do not reveal whether this is a cause of cellular ageing or a consequence of some other macromolecular change.

Ageing of Cells and Populations

It is very important to distinguish between the death of individual cells and of whole populations. It was first pointed out by *Orgel* [49] that all cell lineages are mortal. He considered an experiment in which there is no cellular selection: a cell is allowed to divide, a daughter is picked at random and allowed to divide again, and so on, indefinitely. Eventually a dead cell will be obtained. This cell is not necessarily senescent or aged, it may have died, for instance, as a result of a dominant lethal mutation, the probability of which may be constant with time. Cellular ageing must be defined as a decrease in the probability of survival with time. This is manifested in cell lineages of fibroblasts and also populations where all cell lineages die out. However, it is quite possible to have immortal populations which contain certain ageing cell lineages, but which grow indefinitely as a result of the selection of potentially immortal cells. I will return to this question in a later section.

In a remarkable series of experiments on the longevity of clones of human fibroblasts, *Smith* and his associates [64–66] have demonstrated the heterogeneity of growth potential of individual cells. Moreover, daughters from one individual cell also vary greatly in their proliferative potential, thus demonstrating the stochastic, or non-deterministic nature of the ageing process. In addition, it has been demonstrated with fetal lung strains, MRC-5 and WI-38, that parallel populations (set up from cells stored in liquid nitrogen at early passage) vary considerably in the number of passages or population doublings achieved before growth ceases. *Thompson and Holliday* [69] found that 24 parallel cultures of MRC-5 had an average life span of 57 passages, with a standard deviation of 7.5 passages. (It should be noted that a difference of 10 passages represents 2^{10}, or a 1,000-fold range in growth potential.) It is also clear that the ageing of animals depends at least in part on stochastic processes. Populations of inbred laboratory mice in a uniform environment vary considerably in their longevity. For example, in experimental studies on the ageing of the long-lived CBA strain, *Holliday and Stevens* [23] found that the life span of 35 female control mice was 901 days with a standard deviation of 157 days.

Programmed Ageing

The view that fibroblasts are predetermined to differentiate is, of course, closely related to the hypothesis that the ageing of these cells is programmed. The molecular or cellular basis of such a programme has never been precisely formulated, but presumably a cell division counting mechanism is envisaged: when the clock runs out, cell proliferation ceases. It is not obvious how such a programme could explain the stochastic features of ageing which were previously mentioned, or the strong effect of environmental factors, such as temperature [69]. Nor is it clear why the programme would be altered by mutations which effect, for instance, DNA repair capacity. Most important, it is very hard to explain why a programme for ageing should evolve in the first place [33]. Programmed ageing of cells would not be advantageous and may reduce an organism's fitness, in relation to individuals which do not have such a programme.

Nevertheless, the view that ageing is in some way programmed is widely held, and it is capable, in principle, of explaining the difference in growth potential between diploid and transformed cells, if the 'immortalization' of the latter is due to the destruction or bypass of the programme. Testing the hypothesis is difficult because it has never been formulated in precise terms. Nevertheless, if genes for immortalization exist, then an obvious approach would be to try to isolate them by recently developed methods of genetic manipulation and identify the physiological role of their products.

Genetic Control of Ageing and Its Genetic Consequences

Since animal species have a wide range of maximum longevities, there can be no doubt that ageing is genetically determined. The correlation between the life span of an organism and the growth potential of cultured fibroblasts holds for man, chicken and mouse [13] and probably several other species [56]. However, it is often mistakenly supposed that the genetic control of the rate of ageing in vitro or in vivo, implies that ageing must be programmed. This is incorrect, since genetic factors are responsible for a variety of cellular processes which may influence ageing, including the accuracy of protein synthesis, protein turnover, the efficiency of DNA repair, and the activity of enzymes, such as superoxide dismutase, which remove free radicals.

Mutations which influence the rate of ageing provide a powerful means of investigating primary causes. In man, the rare autosomal recessive Werner's syndrome has a pleiotropic effect in accelerating the onset of a variety of the normal symptoms of ageing, and also greatly reduces the growth potential of cultured skin fibroblasts [40, 72]. Although it is a matter of debate whether the mutation really induces premature ageing in toto [39], it would be very surprising if the identification of the nature of the biochemical or metabolic defect did not provide very important information about the origins of some ageing processes, both in vitro and in vivo.

Although the role of somatic mutations in ageing has frequently been discussed, both theoretical arguments and the experimental evidence strongly suggest that these are more likely to be a consequence rather than a cause of ageing [19]. *Thompson and Holliday* [71] and *Hoen* et al. [15] have shown that tetraploid human fibroblasts have the same longevity as diploid ones, which would not be predicted if deleterious recessive mutants played an important role in ageing. The same conclusions can be drawn from longevity studies of animals of different ploidy [42].

There have been few successful attempts to actually measure the frequency of mutations during ageing. *Fulder and Holliday* [8] screened MRC-5 fibroblast populations for rare variants with a significantly enhanced level of glucose-6-phosphate dehydrogenase, which could have arisen from mutation, and showed that their frequency increased exponentially with culture age [7]. *Morley* et al. [44] measured the frequency of human T-lymphocytes which were resistant to 6-thioguanine from donors of different age, and found a strong correlation between this frequency and age. The data were more consistent with an exponential, than a linear increase. Subsequently, *Morley* et al. [45] have cultured human lymphocytes and have demonstrated that the 6-TG resistant cells are indeed mutants. It had previously been shown that chromosome abnormalities increase sharply in the lymphocytes of old individuals [28] and also in late passage human fibroblasts [59, 70]. Although much more information is needed and better methods are required to screen genetic damage during ageing, the evidence to date rather strongly suggests that the ageing process itself may be mutagenic.

Error Propagation

Orgel [49] first pointed out that the transcription and translation apparatus is potentially unstable. Some errors in the synthesis of proteins may feed back into the highly specific pathway of information transfer from DNA to protein, thereby producing additional errors over and above the intrinsic error level. He later pointed out that, depending on the degree of feedback, errors may simply stabilize at a steady-state level, or they may increase with time, with or without cell division, until a lethal error catastrophe is reached [50]. Although *Hoffman* [16] attempted to demonstrate that error catastrophes were unlikely to occur in biological systems, *Kirkwood and Holliday*

[31] showed that *Hoffman's* [16] model was based on false assumptions and that *Orgel's* [49] original formulation was essentially correct. That is, cells could be in a steady state and therefore potentially immortal, or in a metastable condition, with a given probability of moving into an unstable one. This provides a possible theory of cellular ageing with the following basic features: (1) The initial changes would be cytoplasmic rather than nuclear, since the errors are in proteins. (2) The increase in errors with time would be essentially exponential; thus the observable or measurable phenotypic effects would accumulate rapidly towards the end of the life span. (3) A variety of secondary consequences would be expected from a build up of errors in proteins [51], for example, errors in DNA replication, defects in membranes and imperfect cation transport, progressive abnormalities in organelles, such as mitochondria and lysosomes. (4) The senescent cells should have a dominant effect in the heterokaryon or hybrid, since the 'young' machinery for protein synthesis would become rapidly contaminated with faulty components from senescent cytoplasm. Clear evidence that hybrids selected from young and senescent fibroblasts have very limited growth has recently been obtained by *Pereira-Smith and Smith* [52].

Although the theory of error propagation rests on a sound basis [for a review, see 30], it should be emphasized that it makes no predictions about the level of errors which might be lethal to cells or organisms, and it is possible that the error catastrophe may only affect a small minority of molecules. Nor, in the absence of exact information, is it possible to make meaningful predictions about the rate of increase of errors. It is obviously essential to discover whether error catastrophes can actually occur in biological systems. The first fairly direct evidence came from studies with the mutant leu-5 with *Neurospora crassa*. The phenotype of this strain strongly indicates that it has an altered leucyl tRNA synthetase, with reduced specificity, with the result that it incorporates incorrect amino acids at leucine codons, especially at elevated temperature [54]. The mutant grows continuously at 25 °C, but when shifted to 35 or 37 °C it continues to grow at a constant rate for approximately 3 days and then dies. *Lewis and Holliday* [35] followed the effects on the enzyme glutamic dehydrogenase during this 3-day period and showed that the ratio of inactive cross-reacting material to active enzyme increased dramatically during the last 12 h. The results strongly suggested that the shift in temperature caused an initial increase in protein errors, which was followed 2–3 days later by a secondary effect due to protein error feedback.

In experiments with *Escherichia coli*, conflicting results were initially obtained. *Edelmann and Gallant* [4] and *Gallant and Palmer* [9] claimed to have increased errors in protein synthesis 50-fold by adding streptomycin to the medium, but they observed very little effect on growth and no 'error catastrophe'. On the other hand, *Branscomb and Galas* [2] had previously shown that low concentrations of streptomycin could result in the progressive synthesis of altered β-galactosidase, as judged by its heat lability, together with the cessation of cell division. These experiments have been extended by *Rosenberger* [57], who explored the conditions under which streptomycin affected growth and viability, and has developed, in effect, a model system for studying the finite growth of populations of *E. coli*. At appropriate concentrations of streptomycin, cells continued to grow normally for up to 20 generations, but during this time protein errors increased exponentially, as judged by the suppression of a nonsense codon of β-galactosidase. When growth ceases, there is massive cell death. The fact that populations of *E. coli* can in one environment grow indefinitely, whilst in another they show behaviour very comparable to the Hayflick limit to fibroblast growth, raises the possibility that immortal and mortal populations of animal cells depend on intrinsic differences in error propagation.

Attempts to Measure Error Frequencies in Mammalian Cells

The protein error theory of ageing makes many predictions, which in principle are testable. In practice, it has proved to be very difficult to devise adequate tests. Several papers have been published which purport to disprove the theory, but close examination of the methods used shows that they are inadequate. The basic requirement is to first measure the intrinsic, or spontaneous, level of errors of protein synthesis in normal young cells. Once this is done, the same methods can then be used to measure changes in error levels, if any, during the process of ageing.

Harley et al. [10, 11] and *Wojtyk and Goldstein* [76, 77] have recently claimed that protein errors do not increase during the in vitro ageing of human fibroblasts. In one series of experiments the misincorporation of leucine into polyphenylalanine was measured using cell-free extracts from cultures of different age with poly-U as synthetic message [76, 77]. The percentage of leucine misincorporation (0.03–0.4) was much lower than has been observed in other laboratories, and this was probably due to the addition of suboptimal concentrations of the two amino acids. Indeed, the poly-U system yields widely different error levels with eukaryotic ribosomes, depending on the experimental conditions used [for a review, see 34]. *Wojtyk and Goldstein*

[77] reported that error frequencies actually *decline* during ageing, but as these results are based on an 800-fold range in the rate of incorporation of phenylalanine by cell-free extracts, they cannot be taken seriously. *Buchanan* et al. [3] optimized the experimental conditions and found the amount of leucine misincorporation is about 1% in cell-free extracts from young or senescent fibroblasts. This is, of course, much higher than the in vivo level of protein errors [5], and it is therefore not altogether surprising that any differences between young and old cells would be obscured by what is, in effect, an in vitro artifact. In a second series of experiments, two-dimensional gels were used to look for small subfractions of protein molecules with altered charge [10]. None were seen in protein extracts from senescent cells. Since the method does not detect any errors in proteins from young cells, it is not possible to know whether or not the error level changes in old ones. The authors also measured error levels in cells treated with histidinol. This analogue blocks the normal charging of hystidyl tRNA and from earlier studies with bacterial cells it is known that it results in the misincorporation of wrong amino acids (particularly glutamine) at histidine codons. Under these conditions of histidine starvation, old and young cells produced roughly the same proportion of altered protein, from which the authors conclude that the *spontaneous* level of errors is the same! The theory [11] on which this latter conclusion is based is very simplistic, in that it assumes that the levels of histidyl tRNA in young and old cells are the same under all conditions and also that ribosomes from old cells, if ambiguous, would have increased affinity for non-cognate amino acids but unaltered affinity for cognate amino acids. Neither of these assumptions are justified.

Much stronger evidence against the error theory comes from experiments with viruses. *Holland* et al. [17] infected young and old populations of fibroblasts with vesicular stomatitis, herpes simplex or polioviruses and found no specific differences in yield, heat stability, or, in the case of poliovirus, mutation frequency. This suggests that the pathways of macromolecule information transfer are unimpaired during senescence. However, nothing is known about the effects of cellular errors on these viruses, and it would have been advantageous to find out if treatment of cells with RNA base or amino acid analogues had any measurable effects. *Fulder* [6] examined the reversion frequency of three ts mutants of herpes simplex virus which were grown in young and senescent MRC-5 and found that in one case, reversion frequency was elevated in senescent cells; in another, it was reduced and in a third, it was unchanged. Experiments with virus probes clearly need to be interpreted with great caution.

Considerable indirect evidence in favour of the error theory has accumulated over the last 10 years, particularly from studies with MRC-5 in this laboratory. DNA polymerase α was chosen for detailed investigations of fidelity, because it is a key enzyme in information transfer. It is known that amino acid substitutions reduce the accuracy of DNA polymerase in bacteriophage T4 [48], so a prediction of the theory would be that the enzyme from senescent cells should have reduced accuracy in an appropriate in vitro assay system. This was first tested by *Linn* et al. [37], who found that polymerase α from senescent cells was several times less accurate than the same enzyme from young cells. Much more detailed studies were subsequently carried out by *Murray and Holliday* [47] with enzyme from populations of cells of increasing age. Several template primer systems were used and misincorporation of an incorrect deoxynucleotide triphosphate was measured in the presence of Mg^{++} or Mn^{++}. A variety of control experiments established that the errors seen were indeed due to mistakes made by DNA polymerase α. In addition, evidence was obtained that DNA polymerase γ also lost fidelity during cellular ageing.

In earlier studies, the heat stability of glucose-6-phosphate and 6-phosphogluconate dehydrogenases (G-6-PD and 6-PGD) had been measured in cell-free extracts from MRC-5 cultures, and it was found that late passage cultures always contained a significantly increased fraction of heat-labile enzyme [24]. Although this result has been confirmed by several other laboratories, the interpretation has remained controversial [for a review, see 25]. The results are, however, fully consistent with the possibility that the misincorporation of amino acids increases in senescent cultures, especially as it is known that naturally occurring variants of G-6-PD (new alleles arising from amino acid substitutions) are often heat-labile, and that agents which are known to reduce the fidelity of protein synthesis, such as 5-fluorouracil (5-FU) or paromomycin (Pm) [3], increase the proportion of heat-labile enzyme [22, 24, 25].

Pm has also recently been shown to accelerate many of the normal features of ageing [22]. MRC-5 cells grown in the presence of the antibiotic are unaffected for many generations of growth, but then adopt the morphological characteristics of senescence much sooner than control cultures, including the accumulation of autofluorescent 'age pigments'. Moreover, the long-term effects of Pm are not removed by returning cells to normal medium. It was also shown that as the cells age, they become progressively more sensitive to the effects of Pm. These results, together with earlier experiments with 5-FU, certainly confirm one of the major predictions of the error theory.

Phenotype of Senescent MRC-5 Fibroblasts

The human fetal lung strain MRC-5 was originally characterized by *Jacobs* et al. [27], who showed that it maintains a constant karyotype and rate of growth for 40–50 population doublings. Subsequently, the growth rate slows down and cultures usually die out between 55 and 70 population doublings. It is a striking fact that when cultures are monitored throughout their in vitro life span, significant changes in the phenotype are normally detected only in the last 10–20 population doublings. These experiments are listed in table I, where Y→ O(E) indicates that the result is more compatible with an exponential change in the particular phenotypic characteristic under study, than with a linear increase throughout the life span. This is particularly well-documented for the increase in autofluorescence [55] and variants with high G-6-PD activity [7]. In some cases, only young and senescent cultures were compared (Y and O) and a significant phenotypic difference was observed between them. Experiments with phosphoglucose isomerase produced negative results. *Shakespeare and Buchanan* [62] used immunological methods, which would detect about 15% of inactive cross-reacting material, but none was detected in senescent cultures. However, almost all the other results with MRC-5 either directly support, or are compatible with, the general error theory of cellular ageing.

Unified Theory of Cellular Ageing

In almost all animals, somatic tissues have a finite life span, whereas germ-line lineages are potentially immortal. Although cultured cells are derived from somatic tissue, they differ from those in vivo, because there is ample opportunity for selection of long-lived or potentially immortal populations. The fact that this does not happen in human fibroblast cultures must mean that all the cells are committed to senescence. The commitment theory of cellular ageing was proposed by *Kirkwood and Holliday* [32] and *Holliday* et al. [20] to account for the major biological features of in vitro ageing, and also provide a possible basis for the difference between diploid and transformed populations. Two of these features of ageing are the well-documented variability in life spans of both populations and clones, and the observation that in populations containing two distinguishable cell phenotypes, one frequently becomes predominant during senescence, even in the absence of a selective growth advantage [20, 78]. It was proposed that prior to the

Table I. The phenotype of senescent MRC-5 fibroblasts, based on experiments in the Mill Hill laboratories

Phenotypic characteristics	Cells examined	Reference
1 Increased proportion of heat-labile G-6-PD and 6-PGD	Y → O (E)[1]	*Holliday and Tarrant* [24]
2 Increased proportion of lactic dehydrogenase cross-reacting material	Y → O (E)	*Lewis and Tarrant* [36]
3 Reduced activity and fidelity of DNA polymerase α	Y and O[2]	*Linn* et al. [37]
Reduced activity and fidelity of DNA polymerase α	Y → O (E)	*Murray and Holliday* [47]
Reduced fidelity of DNA polymerase γ		*Murray* [46]
4 Slower rate of replicon elongation	Y and O	*Petes* et al. [53]
5 Increased protein turnover and lysosomal proteolytic activity	Y and O	*Shakespeare and Buchanan* [61, 63]
6 Increased frequency of variants with elevated G-6-PD	Y → O (E)	*Fulder and Holliday* [8] *Fulder* [7]
7 Increased polyploidy and chromosomal abnormalities	Y → O (E)	*Thompson and Holliday* [70]
8 Increased autofluorescence	Y → O (E)	*Rattan* et al. [55]
9 Greater sensitivity to paromomycin	Y → O	*Holliday and Rattan* [22]

[1] Y → O indicates that observations were made at intervals throughout the in vitro life span. (E) indicates that the change seen is more compatible with an exponential than a linear increase. In two experiments (6 and 8), the data very strongly suggest an exponential increase.
[2] Y and O indicates that early and late passage cultures were compared.

establishment of primary fibroblast cultures, diploid cells are potentially immortal, or uncommitted. However, during cell division they give rise with a given probability (P) to committed cells, which continue to divide for many generations, but finally die out. The number of divisions between commitment and death is defined as the incubation period (M) and to facilitate the mathematical treatment this is assumed to be constant. (This model was derived from the experimental evidence for random commitment and a constant incubation period in very detailed studies of the ageing of populations of cells of the fungus, *Podospora* [38]. However, the fibroblast model does not depend on a constant value for M; it could vary considerably in different cell lineages.) It follows that the proportion of uncommitted cells will progressively decline and, if M is sufficiently long, these cells will eventually be lost from the population. It can be shown that if $P \sim 0.25$ and $M \sim 55$ generations, then cultures will inevitably die out. The model makes the surprising prediction that *population size* will influence longevity, and this has been confirmed by 'bottleneck' experiments, in which a transient reduction in the culture size is introduced at various levels [20, 21]. Computer simulations show that the stochastic features of fibroblast ageing and the sorting out of mixed populations are explained by 'random drift' in the loss of the final small number of uncommitted cells. The model is also supported by the observed growth rate of fibroblast populations. A rapid period of proliferation, with very few non-cycling cells, is succeeded by a period of constant slower growth, with approximately 20% non-cycling cells, and finally by senescence and death [21].

In molecular terms, we propose that the potentially immortal diploid cells are in a metastable state, with regard to the stability of their translation apparatus, and have a fairly high probability (~ 0.25) of initiating error propagation. The build up of errors during the incubation period (M) is presumed to be slow and does not kill the cells until an average of ~ 55 generations have elapsed. It is now quite clear that the accurate synthesis of macromolecules does not just depend on enzyme specificity, but also on a range of editing or proof-reading processes, which consume energy [for a review, see 5]. It has been pointed out by *Kirkwood* [29] and *Kirkwood and Holliday* [33] that each species must balance the advantages of the accurate synthesis, maintenance, replacement or repair of macromolecules, against the metabolic cost. The resources necessary for prolonged survival of cells may be better diverted into rapid growth to sexual maturity and reproduction. This will increase overall fitness, but the result is the eventual ageing and death of the soma. Thus the error theory is able to explain evolution of ageing of cells and organisms, whereas other theories of ageing are unable to do so.

Cells will grow indefinitely if the feedback of errors is below a critical level, since then the probability of commitment becomes very low or zero. The theory proposes that germ-line cells are intrinsically more accurate than somatic cells, although it must also be borne in mind that there may be special mechanisms which select out defective cells or lineages [18, 49]. It is possible that transformed cells are also more accurate than diploid somatic cells, perhaps because they are physiologically more similar to proliferating embryonic cells, and can therefore escape from in vitro senescence. However, contrary to the assertion of *Harley* et al. [10], the error theory does not predict that this is necessarily the case. They showed that under conditions of histidine starvation, transformed cells are more error-prone. If transformed cells are also in a metastable state ($p \sim 0.25$), but errors escalate more quickly ($M < 50$ generations), then it is easy to show that populations will contain constant proportions of uncommitted, committed and dead cells, and will grow indefinitely [32].

The unified theory suggests answers to the three major problems raised in the introduction and its predictions can, in principle, be tested experimentally by comparing cells from short- and long-lived species, and somatic and germ-line cells. However, further advances will depend on the development of better methods for measuring accuracy in macromolecule synthesis. At the cellular level, it might be possible to prevent the dilution-out of uncommitted cells and thereby obtain immortal populations of normal diploid cells. It would be even more important to devise experimental conditions which would convert a population of transformed cells with indefinite growth into one which aged and died out. The implications of such a discovery would be far-reaching.

References

1   Bell, E.; Marek, L.F.; Levinstone, D.S.; Merrill, C.; Sher, S.; Young, I.T.; Eden, M.: Loss of division potential in vitro: aging or differentiation? Science *202:* 1158–1163 (1978).
2   Branscomb, E.W.; Galas, D.J.: Progressive decrease in protein synthesis accuracy induced by streptomycin in *E. coli.* Nature, Lond. *254:* 161–163 (1975).
3   Buchanan, J.H.; Bunn, C.L.; Lappin, R.I.; Stevens, A.: Accuracy of in vitro protein synthesis: translation of polyuridylic acid by cell free extracts of human fibroblasts. Mech. Age. Dev. *12:* 339–353 (1980).
4   Edelmann, P.; Gallant, J.: On the translational error theory of ageing. Proc. natn. Acad. Sci. USA *74:* 3396–3398 (1977).
5   Fersht, A.R.: Enzyme editing mechanisms and the genetic code. Proc. R. Soc. *212:* 351–379 (1981).
6   Fulder, S.J.: Spontaneous mutations in ageing human cells: studies using a herpes virus probe. Mech. Age. Dev. *6:* 271–282 (1977).
7   Fulder, S.J.: Somatic mutations and ageing of human cells in culture. Mech. Age. Dev. *10:* 101–115 (1978).

8   Fulder, S.J.; Holliday, R.: A rapid rise in cell variants during the senescence of populations of human fibroblasts. Cell 6: 67–73 (1975).
9   Gallant, J.; Palmer, L.: Error propagation in viable cells. Mech. Age. Dev. 10: 27–38 (1979).
10 Harley, C.B.; Pollard, J.W.; Chamberlain, J.W.; Stanners, C.P.; Goldstein, S.: Protein synthetic errors do not increase during ageing of cultured human fibroblasts. Proc. natn. Acad. Sci. USA 77: 1885–1889 (1980).
11 Harley, C.B.; Pollard, J.W.; Stanners, C.P.; Goldstein, S.: Model for messenger RNA translation during amino acid starvation applied to the calculation of protein synthetic error rates. J. biol. Chem. 256: 10786–10794 (1981).
12 Hayflick, L.: The limited in vitro lifetime of human diploid cell strains. Expl Cell Res. 37: 614–636 (1965).
13 Hayflick, L.: The cellular basis of human ageing; in Finch, Hayflick, Handbook of the biology of aging, pp. 159–186 (Van Nostrand-Reinhold, New York 1977).
14 Hayflick, L.; Moorhead, P.S.: The serial cultivation of human diploid cell strains. Expl Cell Res. 25: 585–621 (1961).
15 Hoen, H.; Gryant, E.M.; Johnston, P.; Norwood, T.H.; Martin, G.M.: Non-selective isolation, stability and longevity of hybrids between normal human somatic cells. Nature, Lond. 258: 608–609 (1975).
16 Hoffman, G.W.: On the origin of the genetic code and the stability of the translation apparatus. J. molec. Biol. 86: 349–362 (1974).
17 Holland, J.J.; Kohne, D.; Doyle, M.F.: Analysis of virus replication in ageing human fibroblast cultures. Nature, Lond. 245: 316–319 (1973).
18 Holliday, R.: Growth and death of diploid and transformed human fibroblasts. Fed. Proc. 34: 51–55 (1975).
19 Holliday, R.; Kirkwood, T.B.L.: Predictions of the somatic mutation and mortalisation theories of cellular ageing are contrary to experimental observations. J. theor. Biol. 93: 627–642 (1981).
20 Holliday, R.; Huschtscha, L.I.; Tarrant, G.M.; Kirkwood, T.B.L.: Testing the commitment theory of cellular ageing. Science 198: 366–372 (1977).
21 Holliday, R.; Huschtscha, L.I.; Kirkwood, T.B.L.: Further evidence for the commitment theory of cellular ageing. Science 213: 1505–1508 (1981).
22 Holliday, R.; Rattan, S.I.S.: Evidence that paromomycin induces premature ageing in human fibroblasts; in Sauer, Cellular ageing; Monogr. devl Biol. 17, pp. 221–233 (Karger, Basel 1984).
23 Holliday, R.; Stevens, A.: The effect of an amino acid analogue p-fluorophenylalanine on longevity of mice. Gerontol. 24: 417–425 (1978).
24 Holliday, R.; Tarrant, G.M.: Altered enzymes in ageing human fibroblasts. Nature, Lond. 238: 26–30 (1972).
25 Holliday, R.; Thompson, K.V.A.: Genetic effects on the longevity of cultured human fibroblasts. III. Correlations with altered glucose-6-phosphate dehydrogenase. Gerontol. 29: 89–96 (1983).
26 Huschtscha, L.I.; Holliday, R.: The limited and unlimited growth of SV40 transformed cells from human diploid MRC-5 fibroblasts. J. Cell Sci. 63: 77–99 (1983).
27 Jacobs, J.P.; Jones, C.M.; Baillie, J.P.: Characteristics of a human diploid cell designated MRC-5. Nature, Lond. 227: 168–170 (1970).
28 Jacobs, P.S.; Brunton, W.; Court Brown, W.M.: Cytogenetic studies of leucocytes on the

general population subjects of ages 65 years and more. Ann. hum. Genet. *27:* 353–365 (1964).
29 Kirkwood, T.B.L.: Evolution of ageing. Nature, Lond. *270:* 301–304 (1977).
30 Kirkwood, T.B.L.: Error propagation in intracellular information transfer. J. theor. Biol. *82:* 363–382 (1980).
31 Kirkwood, T.B.L.; Holliday, R.: The stability of the translational apparatus. J. molec. Biol. *97:* 257–265 (1975).
32 Kirkwood, T.B.L.; Holliday, R.: Commitment to senescence: a model for the finite and infinite growth of diploid and transformed human fibroblast in culture. J. theor. Biol. *53:* 481–496 (1975).
33 Kirkwood, T.B.L.; Holliday, R.: The evolution of ageing and longevity. Proc. R. Soc. *205:* 531–546 (1979).
34 Laughrea, M.: Speed-accuracy relationships during in vitro and in vivo protein biosynthesis. Biochemie *63:* 145–168 (1981).
35 Lewis, C.M.; Holliday, R.: Mistranslation and ageing in *Neurospora.* Nature, Lond. *228:* 877–880 (1970).
36 Lewis, C.M.; Tarrant, G.M.: Error theory and ageing in human diploid fibroblasts. Nature, Lond. *239:* 316–318 (1972).
37 Linn, S.; Kairis, M.; Holliday, R.: Decreased fidelity of DNA polymerase activity isolated from ageing human fibroblasts. Proc. natn. Acad. Sci. USA *73:* 2818–2822 (1976).
38 Marcou, D.: Notion de longévité et nature cytoplasmique due déterminant de la sénescence chez quelques champignons. Annls. Sci. nat. Bot. *12:* 653–764 (1961).
39 Martin, G.M.: Genetic syndromes in men with potential relevance to the pathology of ageing; in Bergsmead, Harrison, Genetic effects on ageing, pp. 5–39 (Liss, New York 1978).
40 Martin, G.M.; Sprague, C.A.; Epstein, C.J.: Replicative lifespan of cultivated human cells: effect of donor's age, tissue and genotype. Lab. Invest. *23:* 86–92 (1970).
41 Martin, G.M.; Sprague, C.A.; Norwood, T.H.; Pendergrass, W.R.: Clonal selection, attenuation and differentiation in an in vitro model of hyperplasia. Am. J. Path. *74:* 137–154 (1974).
42 Maynard-Smith, J.: The causes of ageing. Proc. R. Soc. *157:* 115–127 (1962).
43 Medawar, P.B.: An unsolved problem in biology (Lewis, London 1952); reprinted in 'The Uniqueness of the Individual' (Methuen, London 1957).
44 Morley, A.A.; Cox, S.; Holliday, R.: Human lymphocytes resistant to 6-thioguanine increase with age. Mech. Age. Dev. *19:* 21–26 (1982).
45 Morley, A.A.; Trainor, K.J.; Seshadri, R.; Ryall, R.G.: Measurement of in vivo mutations in human lymphocytes. Nature, Lond. *302:* 155–156 (1983).
46 Murray, V.: Properties of DNA polymerases from young and ageing human fibroblasts. Mech. Age. Dev. *16:* 327–344 (1981).
47 Murray, V.; Holliday, R.: Increased error frequency of DNA polymerases from senescent human fibroblasts. J. molec. Biol. *146:* 55–76 (1981).
48 Muzyczka, N.; Poland, R.L.; Bessman, M.J.: Studies on the biochemical basis of spontaneous mutation. I. A comparison of the deoxyribonucleic acid polymerases of mutator, antimutator, and wild type strains of bacteriophage T4. J. biol. Chem. *247:* 7116–7122 (1970).
49 Orgel, L.E.: The maintenance of the accuracy of protein synthesis and its relevance to ageing. Proc. natn. Acad. Sci. USA *49:* 517–521 (1963).
50 Orgel, L.E.: The maintenance and accuracy of protein synthesis and its relevance to ageing; a correction. Proc. natn. Acad. Sci. USA *67:* 1476 (1970).

51 Orgel, L.E.: Ageing of clones of mammalian cells. Nature, Lond. *243:* 441–445 (1973).
52 Pereira-Smith, O.M.; Smith, J.R.: The phenotype of low proliferative potential is dominant in hybrids of normal human fibroblasts. Somatic Cell Genet. *8:* 731–742 (1982).
53 Petes, T.D.; Farber, R.A.; Tarrant, G.M.; Holliday, R.: Altered rate of DNA replication in ageing human fibroblast cultures. Nature, Lond. *251:* 434–436 (1974).
54 Printz, D.B.; Gross, S.R.: An apparent relationship between mistranslation and an altered leucyl tRNA synthetase in a conditional lethal mutant of *Neurospora* crassa. Genetics *55:* 451–467 (1967).
55 Rattan, S.I.S.; Keeler, K.D.; Buchanan, J.H.; Holliday, R.: Autofluorescence as an index of ageing in human fibroblasts in culture. Biosci. Rep. *2:* 561–567 (1982).
56 Röhme, D.: Evidence for a relationship between longevity of mammalian species and lifespans of normal fibroblasts in vitro and erythrocytes in vivo. Proc. natn. Acad. Sci. USA *74:* 4876–4880 (1981).
57 Rosenberger, R.F.: Streptomycin-induced protein error propagation appears to lead to cell death in *Escherichia coli.* IRCS med. Sci. *10:* 874–875 (1982).
58 Sakagami, H.; Mitsui, Y.; Murota, S.; Yamada, M.: Two-dimensional electrophoretic analysis of nuclear acidic proteins in senescent human diploid cells. Cell Struct. Funct. *4:* 215–225 (1979).
59 Saksela, E.; Moorhead, P.S.: Aneuploidy in the degenerative phase of serial cultivation of human cell strains. Proc. natn. Acad. Sci. USA *50:* 390–395 (1963).
60 Shmookler Reis, R.J.; Goldstein, S.: Loss of reiterated DNA sequences during serial passage of human diploid fibroblasts. Cell *21:* 739–750 (1980).
61 Shakespeare, V.; Buchanan, J.H.: Increased degradation rates of protein in ageing human fibroblasts and in cells treated with an amino acid analog. Expl Cell Res. *100:* 1–8 (1976).
62 Shakespeare, V.; Buchanan, J.H.: Studies on phosphoglucose isomerase from cultured human fibroblasts: absence of detectable ageing effects on the enzyme. J. cell. Physiol. *94:* 105–116 (1978).
63 Shakespeare, V.A.; Buchanan, J.H.: Increased proteolytic activity in ageing human fibroblasts. Gerontol. *25:* 305–313 (1979).
64 Smith, J.R.; Hayflick, L.: Variation in the life-span of clones derived from human diploid cell strains. J. Cell Biol. *62:* 48–53 (1974).
65 Smith, J.R.; Whitney, R.G.: Intraclonal variation in proliferative potential of human diploid fibroblasts: stochastic mechanism for cellular ageing. Science *207:* 82–84 (1980).
66 Smith, J.R.; Pereira-Smith, O.M.; Schneider, E.L.: Colony size distribution as a measure of in vivo and in vitro aging. Proc. natn. Acad. Sci. USA *75:* 1353–1356 (1978).
67 Sun, T.T.; Green, H.: Differentiation of the epidermal keratinocyte in cell culture: formation of the cornified envelope. Cell *9:* 511–521 (1976).
68 Swim, H.A.; Parker, R.F.: Culture characteristics of human fibroblasts propagated serially. Am. J. Hyg. *66:* 235–243 (1957).
69 Thompson, K.V.A.; Holliday, R.: Effect of temperature on the longevity of human fibroblasts in culture. Expl Cell Res. *80:* 354–360 (1973).
70 Thompson, K.V.A.; Holliday, R.: Chromosome changes during the in vitro ageing of MRC-5 human fibroblasts. Expl Cell Res. *96:* 1–6 (1975).
71 Thompson, K.V.A.; Holliday, R.: The longevity of diploid and polyploid human fibroblasts: evidence against the somatic mutation theory of cellular ageing. Expl Cell Res. *112:* 28–287 (1978).

72 Thompson, K.V.A.; Holliday, R.: Genetic effects on the longevity of cultured human fibroblasts. I. Werner's syndrome. Gerontol. *29:* 73–82 (1983).
73 Thompson, K.V.A.; Holliday, R.: Genetic effects on the longevity of cultured human fibroblasts. II. DNA repair deficient syndromes. Gerontol. *29:* 83–88 (1983).
74 Viermy, C.; Keller, A.M.; Begel, O.; Belcour, L.: A sequence of mit DNA is associated with the onset of senescence in a fungus. Nature, Lond. *297:* 157–159 (1982).
75 Witkowski, J.A.: Dr. Carrel's immortal cells. Med. Hist. *24:* 129–142 (1980).
76 Wojtyk, R.I.; Goldstein, S.: Fidelity of protein synthesis does not decline during ageing of cultured human fibroblasts. J. cell. Physiol. *103:* 299–303 (1980).
77 Wojtyk, R.I.; Goldstein, S.: Clonal selection in cultured human fibroblasts: role of protein synthetic errors. J. Cell Biol. *95:* 704–710 (1982).
78 Zavala, C.; Fialkow, P.J.; Herner, G.: Evidence for selection in cultured diploid fibroblast strains. Expl Cell Res. *117:* 137–144 (1978).

R. Holliday, PhD, FRS, National Institute for Medical Research, The Ridgeway, Mill Hill, London NW7 1AA (England)

II. In vitro vs. in vivo Ageing

Cellular Aging of Human Retinal Epithelium in vivo and in vitro[1]

M.T. Flood, J.E. Haley, P. Gouras[2]

Columbia University, Department of Ophthalmology, New York, N.Y., USA

Introduction

The human retinal pigment epithelium (RPE) is a single layer of pigmented epithelial cells that surrounds the outer circumference of the retina from the optic nerve to the ora serrata and rests on a basal lamina that is continuous with the inner lamella of Bruch's membrane of the choroid. The apical surface of these cells is covered with many microvilli processes which extend outward between the outer segments of the photoreceptors and actually clasp the photoreceptor outer segments. In addition to supplying many of the metabolic needs of the neural retina, the RPE cells are the major storage depot for vitamin A used in vision [4] and are responsible for phagocytizing the discarded outer segment discs of the photoreceptor cells [3, 30]. The functional and metabolic relationship between the retinal epithelium and the photoreceptors [1, 30] and between the retinal epithelium and Bruch's membrane [7, 8, 17, 21] as well as abnormalities of the retinal epithelium observed in aging [23] have focused attention on the vulnerability of this cell layer in the aging process.

The RPE cells are particularly interesting for cellular aging studies. First of all, these cells do not undergo division after fetal development in vivo [28] so that popular theories of aging based on mitotic activity do not apply to these cells. Secondly, these cells contain two types of age-related pigment

---

[1] Supported by NEI Grant EY 03854 from NIH, by the National Retinitis Pigmentosa Foundation and by the House of St. Giles the Cripple.

[2] The authors wish to thank *Hild Kjeldbye* for assistance with electron microscopy, *Mary Bilek* and *Eda Maiello* for technical assistance, and *Anne Leitch* for secretarial services.

granules: one type is melanin which is synthesized during fetal development [5, 9, 27, 28] and is usually found in and near the apical processes of the cell where it functions as a sink for incoming light that transverses the retina; the second type of pigment granule is lipofuscin, a degradation product mainly of the phagocytized outer segments which accumulates with age [9, 10, 22]. RPE cells are also known to undergo pleomorphic changes with growth and aging [15, 27, 28] and occasional drusenoid inclusions in older eyes cause a greater separation of aging RPE cells from the choriocapillaris [11].

Because of the delicate nature of the human retina as well as the importance of each part of the retina in normal visual function, biopsy material for cellular aging studies on the human retina has been very limited. The postmortem viability of human RPE cells and the ability to maintain these cells in tissue culture for extended periods of time has opened new possibilities in human aging research.

Materials and Methods

Donor Eyes

Postmortem human donor eyes were received through the New York Eye Bank for Sight Restoration, the donor eye program of the National Retinitis Pigmentosa Foundation, and the National Diabetes Research Interchange. The eyes ranged in chronological age from 7 days to 100 years and were generally received within 24–40 h post mortem.

Cell Culture

Primary Cell Cultures. Primary RPE cell cultures were established from donor eyes received within 40 h post mortem. The eyes were processed in a sterile environment according to the method previously published [12]. In the procedure, the cornea and an additional 5-mm-wide band of sclera at the corneal-scleral junction were removed first. An aliquot of liquid vitreous was then aspirated through a 20-gauge needle inserted through the ora serrata into the liquid pocket of the vitreous. A circular cut was made through the choroid, retinal epithelium and neural retina, following the circumference of the scleral rim, and the anterior segment with adherent vitreous was lifted out of the eyecup. The neural retina was separated from the optic disc and removed. The shell was then washed with Hanks' balanced salt solution, Ca^{++}- and Mg^{++}-free, and treated with 0.25% trypsin (Flow Laboratories) for 1–2 h at 37 °C. The trypsin solution was aspirated from the shell and Eagle's minimum essential medium (MEM) (Flow Laboratories) was added. The cells were released from Bruch's membrane by gently pipetting the culture medium in the shell. The concentration of cells in the suspension was determined with a hemocytometer and 2 ml of cell suspension containing 1 to 3×10^5 cells/ml was plated out in Corning polystyrene Petri dishes (35 mm in diameter). The cultures were incubated in Eagle's MEM plus 20% fetal calf serum and 0.5% glucose in a moist chamber at 37 °C in 5% CO_2 and 95% air atmosphere. 3–10 days were allowed for initial attachment. After the cells attached, the culture medium was changed every 3–4 days.

Subcultures. Subculturing was done when the primary cultures became confluent. A solution of 0.05% trypsin and Versene (1:1,000) was added to the Petri dishes for 5–7 min at 37 °C to remove the cells. The trypsin-Versene solution was removed after centrifugation and the cells were plated out in Corning Petri dishes at concentrations of 5 to 10×10^4 cells per culture plate.

Macular and Peripheral RPE Cell Cultures. The RPE cells of the macular area and specific peripheral areas of the retina were isolated after trypsin incubation. A keratome was placed over the macular or peripheral area (encompassing an area 8 mm in diameter) and the cells were removed by pipetting the culture medium within the keratome. In each case the cells were maintained in MEM plus 20% fetal calf serum and 0.5% glucose.

Growth Patterns. The cell cultures were observed daily with an inverted phase microscope and the number of cells per unit area was determined in different sections of the culture plate. The growth patterns were analyzed with respect to the cell concentration in the initial inoculum and the number of cells that attached. The influence of the chronological age of the donor, the postmortem time, and relevant clinical history on the in vitro growth characteristics were studied.

Two-Dimensional Gel Electrophoresis

Protein components of primary and subcultured cells were studied by two-dimensional gel electrophoresis. Confluent primary cultures and subcultures were washed with 2 ml of methionine-free Eagle's MEM (Flow Laboratories) containing 20% fetal calf serum. This medium was removed and replaced with 1 ml of the same medium containing 140 μCi (^{35}S)-*L*-methionine (New England Nuclear) (specific activity 935.7 Ci/mM) and the cultures were incubated for 6 h. 6 h was the optimum steady-state labeling time to produce proteins with a total specific activity of 9 to 20×10^7 dpm/mg protein. The cells were then washed three times with phosphate-buffered saline, pH 7.4, removed from the culture plate surface with a rubber policeman and centrifuged at 1,200 g for 10 min. The pelleted cells were prepared for two-dimensional electrophoresis as described by *Garrells* [16]. This procedure included treatment with DNase-RNase for 30 min in the presence of sodium dodecyl sulfate (SDS) at 4 °C and lyophilization and subsequent solubilization with sonication in buffer containing 9.95 M urea, 4% NP-40, 2% ampholytes (pH 6–8, LKB) and 100 mM dithiothreitol. The final SDS concentration was 0.3%. About 500,000 cpm of radioactivity was loaded onto the isoelectric focusing gels. The conditions for isoelectric focusing, molecular weight separation, and fluorography of the dried gels in the second dimension have been previously described [18, 19].

Light and Electron Microscopy

Pieces of retinal epithelium (1 mm^2) with attached choroid and sclera were fixed in 3% glutaraldehyde at 4 °C for 2 h to study the in vivo morphology of aging. The tissue was then washed in Earle's buffer and postfixed in 1% osmium tetroxide for 1 h. After acetone dehydration, the tissue was embedded in Epon, polymerized, and sectioned. Sections (1 μm thick) were stained with toluidine blue for light microscopy. Thin sections were also cut, stained with uranyl acetate and lead citrate, and studied on a Siemens Elmiskop to correlate in vivo ultrastructure with in vitro growth properties.

Confluent cell cultures were also processed for electron microscopy. In these preparations, the culture medium was removed and the cultures were fixed in 3% glutaraldehyde at 4 °C for 2 h. The cultures were then washed with Earle's buffer and postfixed in 1% osmium tetroxide for 1 h. After ethanol dehydration, Epon was poured into the culture plates and polymerized.

The cultures were sectioned perpendicularly to the culture surface to study monolayer growth and basal lamina formation. Thin sections were stained with uranyl acetate and lead citrate and examined on a Siemens Elmiskop.

Results

Electron Microscopy Studies in vivo

Ultrastructural studies of RPE obtained from young and older donor eyes manifested morphological changes that occur in these cells with aging. Figure 1 shows retinal epithelium, Bruch's membrane, and choroidal tissue from a 7-day-old donor eye. The RPE cells in young eyes characteristically contain a single type of pigment granule, melanin, which is elliptical in shape and distinguishable from lipofuscin by its homogeneous density. In contrast, RPE cells from a 58-year-old donor (fig. 2) contain mostly lipofuscin and melanolipofuscin granules; in these cells melanin granules are rare. The amount of lipofuscin increases with increasing chronological age [9, 10].

Bruch's membrane also shows age-related changes that distinguish young and older eyes. In young eyes, Bruch's membrane is usually a uniform

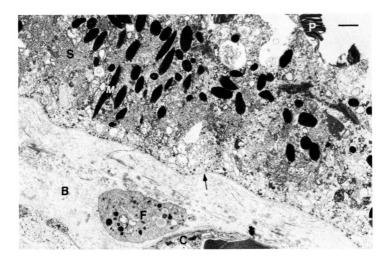

Fig. 1. Transmission electron micrograph of human retinal epithelium and Bruch's membrane from a 7-day-old donor. Note the melanin granules (M), smooth endoplasmic reticulum (S), and the thin basal lamina (↑) of the RPE cell. Bruch's membrane (B) consists of a loose collection of collagen and elastin fibers and contains a fibroblast cell (F). P = Photoreceptor outer segment; C = choriocapillaris. Bar = 1 μm.

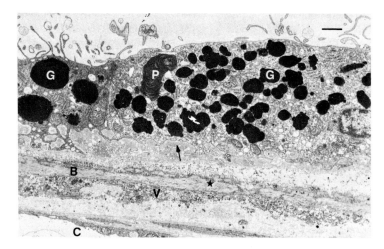

Fig. 2. Transmission electron micrograph of human retinal epithelium and Bruch's membrane from a 58-year-old donor. The pigment granules (G) are lipofuscin or melano-lipofuscin complexes and there is a marked thickening of the basal lamina (↑) of the RPE cell. Bruch's membrane (B) contains arrays of vesicles (V) and fibrous structures (★) in the mid-zone. P = Phagosome; C = lumen of choriocapillaris. Bar = 1 μm.

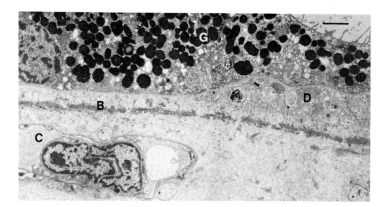

Fig. 3. Transmission electron micrograph of human retinal epithelium and Bruch's membrane (B) from a 65-year-old donor. Note the drusenoid inclusion (D) under the RPE cell and the accompanying thinning of the cell. Pigment granules (G) are lipofuscin. C = Choriocapillaris. Bar = 1 μm.

layer of collagen and elastin fibers, occasionally invaded by a fibroblast cell, and is in continuity with the basal lamina of the RPE cell and the basal lamina of the endothelial cells of the choriocapillaris (fig. 1). The collagen and elastin fibers are interwoven into bundles of fibers in longitudinal and cross-sectional arrays. Fibroblast cells were not observed in Bruch's membrane in nonpathological tissues from donors older than 2 years of age.

In older eyes, that is, eyes from donors over 50 years of age, the fibrous structural arrangement of Bruch's membrane was disturbed (fig. 2). Older eyes characteristically have granular, vesicular and filamentous inclusions randomly distributed throughout Bruch's membrane. In some specimens, there were accumulations of amorphous material and elongated, noncollagen-like fibers in this layer. Drusenoid inclusions were seen in tissue obtained from a 65-year-old donor (fig. 3); the origin of these accumulations under the RPE cells was not ascertained.

Cell Cultures

RPE cells obtained from human donor eyes within 40 h post mortem were generally still viable and were able to be maintained in culture. Electron microscopy studies of RPE cells in culture showed that the cells in primary culture retained many of their in vivo properties (fig. 4). The cells typically had an apical-basal polarity, many microvilli projections on the apical cell surface, dense infoldings of the basal cell surface, pigment granules, and various types of junctional complexes.

Daily observation with a phase contrast microscope and counting of the cells in culture revealed two trends: first, that there was an age-dependent control of the onset of division in these cell cultures, and secondly, that the number of cells that were potential dividers decreased with increasing chronological age.

In each viable culture only a percentage of the cells went into division and gave rise to confluency; the other cells remained stationary, that is, they did not go into division and remained heavily pigmented. The cells that went into division lost their pigmentation and became transparent (fig. 4). Therefore, two populations of cells, dividing cells and stationary cells, were easily distinguished from each other in cell cultures.

Daily cell counts in several areas of each culture demonstrated a slow onset of proliferation in RPE cell cultures that was directly related to the chronological age of the donor (fig. 5). The cultures generally remained in an initial lag phase from 3 to 21 days after plating. Following this initial lag phase, the cells either entered into a rapid division phase, increased in

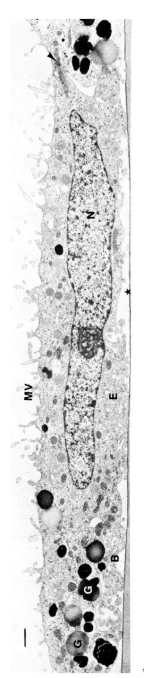
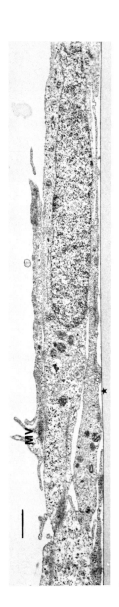

Fig. 4. a Transmission electron micrograph of human RPE cell in primary culture. Note the apical-basal polarity, the microvilli (MV) projections on the apical surface, the deep infoldings of the basal surface (B), the extracellular material (E) on the basal surface, and the junctional complex (arrowhead). The pigment granules (G) are lipofuscin. N = Nucleus; ★ = culture surface. Bar = 1 μm. *b* Transmission electron micrograph of human RPE cell in subculture. Note the apical-basal polarity, the microvilli projections (MV) on the apical surface, and the absence of pigment granules. ★ = Culture surface. Bar = 1 μm.

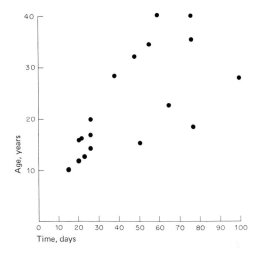

Fig. 5. The relationship of donor chronological age and the time to reach confluency (200 cells/mm^2) in RPE cells in primary culture.

number, and became confluent or they remained stationary. In stationary cultures the cells eventually rounded up and sloughed off the culture surface. The duration of the initial lag phase was age-dependent: cultures from donors less than 10 years of age went into division by day 3; cultures established from donors 10–20 years of age went into division by day 7, and cultures from donors over 30 generally took 14–21 days before division occurred. Some cultures, especially those from older donor eyes, never entered the proliferative stage; they remained viable, but stationary, for long periods of time.

In order to determine what percentage of cells actually entered into division and gave rise to confluency, the number of pigmented cells in several areas of a culture plate was counted prior to the onset of division and compared with the number of pigmented cells in the same areas at confluency. The cultures used in this study were established from donors who ranged in chronological age from 20 to 100 years. In counting 40 separate areas from RPE cell cultures obtained from 12 different donors, we determined that the percent decrease in the number of pigmented cells for all the cells studied was 28% (table I). When the data were analyzed with respect to the chronological age of the donor, the percent decrease for the number of pigmented cells obtained from donors over 50 years of age was approximately 20%; for cells from donors less than 30 years of age, it was about 35%. The data indicated that the proliferative population was small and decreased with increasing

Table I. Comparison of the number of pigmented cells at the onset of the growth phase and at confluency [from ref. 13]

Donor age, years	Number of areas	% decrease
20–100	40	28.1 ± 3.8 (SEM)
>50	18	19.6 ± 3.7 (SEM)
<30	22	34.9 ± 5.9 (SEM)

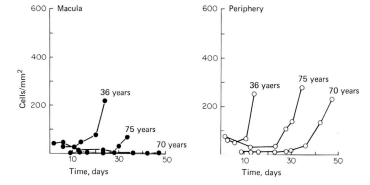

Fig. 6. Growth curves of RPE cell cultures established from the macular or peripheral retinal area from 3 donors. Note the later onset of proliferation or the absence of proliferation in the cells of the macular area. The number beside each curve represents the donor's age.

chronological age. The delay in achieving confluency in cell cultures from older donors appeared to be influenced by the fact that RPE cells from older eyes have a smaller potential proliferative pool.

The decreased proliferative pool may depend upon the amount of lipofuscin in the cells. To test this hypothesis the onset of division was studied in RPE cell cultures established from the macular area and compared with RPE cell cultures established from peripheral retinal area of the same donor eye. Macular cells are known to contain more lipofuscin material than peripheral cells [29]. In RPE cells from each of three pairs of donor eyes studied, the cells from the peripheral areas entered the proliferative phase earlier than the cells from the macular area (fig. 6). In fact, in cultures established from

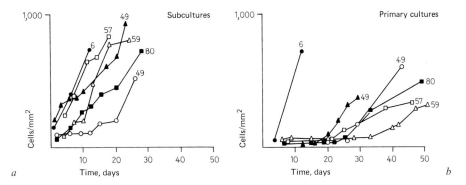

Fig. 7. Growth curves of RPE cells from the same donors in primary cultures *(a)* and subcultures *(b)*. Subcultures were established from confluent primary cultures. The number beside each curve represents the donor's age.

RPE cells of one 70-year-old donor, the cells from the macular area remained stationary for approximately 45 days.

The onset of proliferation in subcultured RPE cells which are devoid of lipofuscin material is rapid and apparently no longer influenced by the chronological age of the donor. Figure 7 shows the growth curves of cells from the same donor cell populations in primary culture and subculture. Note the absence of an initial age-dependent lag phase in the onset of division in the subcultured cells.

Protein Patterns

After 6 h of labeling, the two-dimensional fluorograms of confluent primary and subcultured human RPE cells showed approximately 200 proteins, mostly acidic. At this level of sensitivity the protein patterns of primary and subcultured RPE cells was qualitatively similar (fig. 8). Two-dimensional patterns of proteins obtained from human fibroblasts in culture show many striking differences in their protein patterns from RPE cells [18]. The cytoskeletal proteins of the cell, including actin, and intermediate-sized filament proteins are major proteins common to primary and subcultured cells that have been identified [19]. Although it has not been rigorously quantitated, there appeared to be minor shifts in the relative intensities of some of these proteins. However, there were no apparent differences in the makeup of major proteins in primary and subcultured human RPE cells.

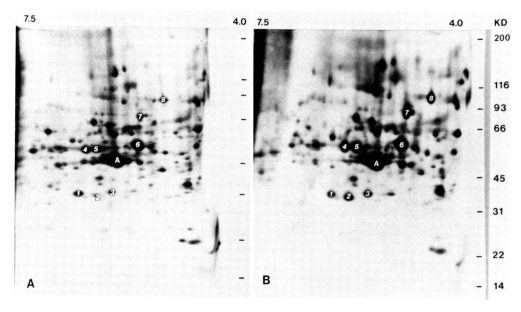

Fig. 8. Protein components of human RPE cells in primary cultures *(A)* and subcultures *(B)* determined by two-dimensional gel electrophoresis. Proteins 1–8 are unique to RPE cells as compared with fibroblasts; No. 4, 5, and 6 are intermediate filament proteins as determined by a cytoskeletal preparation, and A is actin as determined by coelectrophoresis with chicken muscle actin.

Discussion

Human RPE cells in culture represent a unique system for studying aging. First of all, the cells used in this study cover a large chronological age range (7 days to 100 years), in which time they accumulate enormous amounts of lipofuscin. Secondly, these cells are nonreplicators, therefore, aging hypotheses dependent on repeated cell division are not relevant in these experiments. Despite the fact that RPE cells do not divide postnatally, there is an age-dependent loss of proliferative potential and an age-dependent delay in the onset of proliferation in these cells. There must be a biological clock independent of cell division which moves this cell population towards senescence and which must be different from the mitotic clock which limits the life span of fibroblasts in vitro [6, 20, 24, 26].

Our results imply that lipofuscin may be the clock. The role of accumulated lipofuscin in the aging process is currently uncertain. There has been no proof that lipofuscin is a causal factor of aging rather than an effect. Our results suggest that increasing amounts of lipofuscin can have an effect on cell function that becomes manifest when these cells are cultured. The amount of lipofuscin in these cells increases progressively from birth, primarily but not exclusively, as degradation products of the phagocytized rod and cone outer segments that accumulate in the cells [25]. Cells from older donors therefore have more lipofuscin material than cells from younger donors [9]. Cells from the macular area also have more lipofuscin than cells from the peripheral retina [29]. In both cases the cells with the greater accumulation of cellular waste material, lipofuscin, take longer to go into proliferation.

This hypothesis is supported by the growth patterns of subcultured cells which are biochemically similar to the cells in primary culture as shown by two-dimensional gel electrophoretic patterns (fig. 8) and vitamin A studies which we have reported [14]. Since lipofuscin accumulates in the cells from phagocytosis of photoreceptor outer segment material, cells in culture will not increase their store of lipofuscin. As cells divide, the lipofuscin material is distributed and consequently diluted among the daughter cells. Subcultured cells, therefore, have negligible amounts of this cellular debris. Observation of growth curves in second-passage subcultured cells shows a uniform pattern of proliferation with a rapid onset and steady rate of division and no age-dependent initial lag phase. It appears, therefore, that the cells can be rejuvenated by diluting their age-dependent waste product, lipofuscin.

References

1 Bairati, A.; Orzalesi, N.: The ultrastructure of the pigment epithelium and of the photoreceptor pigment epithelium junction in the human retina. J. Ultrastruct. Res. *9:* 484–496 (1963).
2 Bell, E.; Marek, L.F.; Levenstone, D.S.; Merrill, C.; Sher, T.; Young, I.T.; Eden, M.: Loss of division potential in vitro: aging or differentiation. Science *202:* 1158–1163 (1978).
3 Bok, D.; Young, R.W.: Phagocytic properties of the retinal pigment epithelium; in Zinn, Marmor, The retinal pigment epithelium, pp. 148–174 (Harvard University Press, Cambridge 1979).
4 Bridges, C.D.B.; Alvarez, R.A.; Fong, S.I.: Vitamin A in human eyes: amount, distribution, and composition. Investve Ophthal. vis. Sci. *22:* 706–714 (1982).
5 Breathnack, A.S.; Wyllie, L.M.A.: Ultrastructure of retinal pigment epithelium of the human fetus. J. Ultrastruct. Res. *16:* 584–597 (1966).
6 Cristafalo, V.J.; Sharf, B.B.: Cellular senescence and DNA synthesis (thymidine incorporation as a measure of population age in human diploid cells). Expl Cell Res. *76:* 419–427 (1973).

7 Farkas, T.G.; Sylvester, V.; Archer, D.; Altona, M.: The histochemistry of drusen. Am. J. Ophthal. 71: 1206–1215 (1971).
8 Farkas, T.G.; Krill, A.E.; Sylvester, V.M.; Archer, D.: Familial and secondary drusen: histologic and functional correlations. Trans. Am. Acad. Ophthal. Oto-lar. 75: 333–343 (1971).
9 Feeney, L.: Lipofuscin and melanin of human retinal pigment epithelium. Investve Ophthal. vis. Sci. 17: 583–600 (1978).
10 Feeney, L.; Grieshaber, J.A.; Hogan, M.J.: Studies on human ocular pigment; in Rohen, Eye Structure II Symposium, pp. 535–548 (Schattauer, Stuttgart 1965).
11 Fine, B.S.; Kwapien, R.P.: Pigment epithelial windows and drusen: an animal model. Investve Ophthal. vis. Sci. 17: 1059–1068 (1978).
12 Flood, M.T.; Gouras, P.; Kjeldbye, H.: Growth characteristics and ultrastructure of human retinal pigment epithelium in vitro. Investve Ophthal. vis. Sci. 19: 1309–1320 (1980).
13 Flood, M.T.; Gouras, P.: The organization of human retinal epithelium in vitro. Vision Res. 21: 119–126 (1981).
14 Flood, M.T.; Gouras, P.; Haley, J.E.; Blaner, W.S.: Human retinal pigment epithelium in vitro: organization, ultrastructure, and biochemistry; in Cotlier, Maumenee, Berman, Genetic eye diseases: retinitis pigmentosa and other inherited eye disorders, pp. 53–66 (Liss, New York 1982).
15 Friedman, E.; Tso, M.O.M.: The retinal pigment epithelium. II. Histologic changes associated with age. Archs Ophthal. 79: 315–320 (1982).
16 Garrells, J.I.: Two-dimensional gel electrophoresis and computer analysis of proteins synthesized by clonal cell lines. J. biol. Chem. 254: 7961–7977 (1979).
17 Gass, J.D.M.: Drusen and disciform macular detachment and degeneration. Archs Ophthal. 90: 206–217 (1973).
18 Haley, J.E.; Flood, M.T.; Gouras, P.; Kjeldbye, H.M.: Proteins from human retinal pigment epithelial cells: evidence that a major protein is actin. Investve Ophthal. vis. Sci. 24: 803–811 (1983).
19 Haley, J.E.; Flood, M.T.; Kjeldbye, H.; Maielo, E.; Bilek, M.K.; Gouras, P.: Two-dimensional electrophoresis of proteins in human retinal pigment epithelial cells: identification of cytoskeletal proteins. Electrophoresis 4: 133–137 (1983).
20 Hayflick, L.; Moorehead, P.S.: The serial cultivation of human diploid cell stains. Expl Cell Res. 25: 585–621 (1961).
21 Hogan, M.J.: Role of the retinal pigment epithelium in macular disease. Trans. Am. Acad. Ophthal. Oto-lar. 76: 64–80 (1972).
22 Kolb, H.; Gouras, P.: Electron microscopic observations of human retinitis pigmentosa, dominantly inherited. Investve Ophth. 13: 487–498 (1974).
23 Marmor, M.F.: Dystrophies of the retinal pigment epithelium; in Zinn, Marmor, The retinal pigment epithelium, pp. 424–453 (Harvard University Press, Cambridge 1979).
24 Martin, G.M.; Sprague, C.A.; Epstein, C.J.: Replicative life-span of cultivated human cells: effects of donor's age, tissue, and genotype. Lab. Invest. 23: 86–92 (1970).
25 Robison, W.G., Jr.; Kuwabara, T.; Bieri, J.G.: Deficiencies of vitamin E and A in the rat: retinal damage and lipofuscin accumulation. Investve Ophthal. vis. Sci. 19: 1030–1037 (1980).
26 Thrasher, J.D.: Age and the cell cycle of the mouse esophageal epithelium. Exp. Gerontol. 6: 19–24 (1971).

27 Tso, M.O.M.; Friedman, E.: The retinal pigment epithelium. I. Comparative histology. Archs Ophthal. *78:* 641–649 (1967).
28 Tso, M.O.M.; Friedman, E.: The retinal pigment epithelium. III. Growth and development. Archs Ophthal. *80:* 214–216 (1968).
29 Wing, G.L.; Blanchard, G.C.; Weiter, J.L.: The topography and age relationship of lipofuscin concentration in the retinal pigment epithelium. Investve Ophthal. vis. Sci. *17:* 601–607 (1978).
30 Young, R.W.; Bok, D.: Autoradiographic studies on the metabolism of the retinal pigment epithelium. Investve Ophth. *9:* 524–536 (1970).

M.T. Flood, PhD, Columbia University, Department of Ophthalmology,
630 West 168 Street, New York, NY 10032 (USA)

Growth Potential, Repair Capacity and Protein Synthesis in Lens Epithelial Cells during Aging in vitro

Hermann Rink

Institute of Radiobiology, University of Bonn, FRG

Introduction

Early reports of *Carrel* [1912] led to the hypothesis of the immortality of mammalian cells when cultured as isolated cells in adequate media. In other words, the well-known phenomenon of aging of a mammalian organism is reduced to the interaction of all his constituents. This concept has been modified fundamentally, and *Bayreuther* [1975] offered a precise description of the actual version: 'The aging of an organism has to be reduced to the aging of its organs. The aging of the different organs has to be reduced to the aging of their cells.'

The aging behavior of isolated cells can be traced in cell cultures. The basic work was initiated by *Hayflick and Moorhead* [1961] and *Hayflick* [1965]. They were the first to point out that human diploid fibroblasts of embryonic origin proliferate when subcultured serially, but undergo only a distinct number of population doublings after which proliferation stops. With additional transfers the cells degenerate and die [phase III phenomenon according to *Hayflick*, 1965]. Today, we know that the degeneration is characterized by a loss of the proliferative capacity; without further cell transfer, however, these old passage cells maintain their energy and synthesis metabolism over months [*Bayreuther*, 1982]. The loss of proliferative potential, which is the causative factor of the limited lifetime of isolated diploid mammalian cells first described by *Hayflick* [1965], is called cellular aging in vitro.

Fibroblast systems have been widely used to study the aging behavior in vitro, but there are only a few notes on the aging behavior of epithelial cell systems. Pure epithelial cell populations, although only in small amounts, can

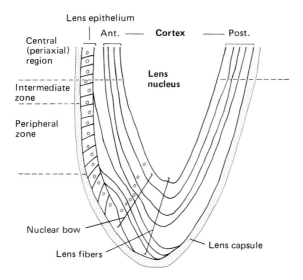

Fig. 1. Schematic diagram of a meridional section through the lens [redrawn from *Wanko and Gavin*, 1959].

be obtained from the crystalline lens. For two reasons the in vitro study of lens epithelial cells is of special interest: (i) comparison to fibroblast systems is necessary, and (ii) aging of lens epithelial cells in situ may contribute to the clinical manifestation of lens opacification in old age.

The lens is an avascular organ, which shows two remarkable age-related peculiarities [*Hockwin*, 1971]. During the whole life span of an individual the lens increases in mass and volume. This unique behavior results from the fine architecture of the lens and the proliferation behavior of its cells. The epithelial cells of lens represent a monolayer of cuboidal cells beneath the anterior lens capsule (fig. 1). Epithelial cells of the central region are mitotically inactive, whereas cells of the peripheral zone show intense mitotic activity. In the equatorial region the cells elongate and differentiate to fiber cells, which later on lose their nuclei. These fiber cells form layer upon layer over the already existing lens mass. Due to this kind of appositional growth the lens never sheds any cells. Cells of the embryonic phase later form the lens nucleus, younger cells constitute the outer layer, the lens cortex.

The lens is rich in proteins, most of them water-soluble and belonging to the lens-specific proteins, called crystallins (α-, β- and γ-crystallins), the composition of which is subject to aging. While the α- and β-crystallins are

Table I. Diploid in vitro life span of lens epithelial cells from different species

Species	Donor age	Life span
Mouse	4–6 days	12–16 CPD[1]
Rat	4–6 days	32–36 CPD[1]
Rat (cat$^+$)	4–6 days	3–5 CPD
Pig	0.7 years	36–42 CPD
Bovine	0.5 years	64–70 CPD[1]

[1] Transformation into aneuploid lines.

present in the epithelial cell in situ, the synthesis of γ-crystallin is related to the differentiation process; it represents a biochemical marker of differentiation [*Papaconstantinou*, 1967].

With increasing age the initially clear lens may lose its transparency. Lens opacities of different types are more frequent in old age. These findings led to the supposition that the phenomenon is correlated with the aging process (cataracta senilis). The exact causes for lens opacification have not yet been elucidated; presumably different mechanisms are involved. Nuclear opacities in the oldest part of the lens nucleus are obviously induced by post-translational protein modifications, which in turn lead to protein aggregations. Cortical and subcapsular posterior opacifications, however, may be due to disturbances in the proliferation and differentiation behavior of lens epithelial cells. Under these aspects it is of crucial importance to investigate lens epithelial cells during aging in vitro, that is, during serial subcultivation up to the end of their diploid in vitro life span and to relate the results to the possible mechanisms which may affect lens transparency in older age.

Results

First studies demonstrated that within the system of lens epithelial cells a relationship exists between the maximum life span of a species and the maximum number of cumulative population doublings (CPD) of its cells in culture systems (table I). An inverse relationship exists between the donor age and the maximal CPD number. This is illustrated in figure 2, showing that rat lens epithelial cells derived from inbred Sprague-Dawley rats (5 days of age) perform 35 CPD before they undergo transformation to aneuploid

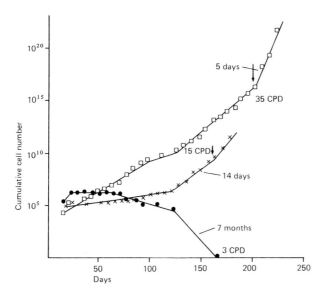

Fig. 2. Growth behavior of rat lens epithelial cells in relation to donor age. Arrows indicate transformation into aneuploid lines. Donor age: □ = 5 days; × = 14 days; ● = 7 months. [From *Vornhagen and Rink*, 1979].

lines [*Rink* et al., 1980]. Cells from 14-day-old donors perform only 15 CPD, and cells from older animals (7 months) only reach 3 CPD before they enter phase III and die with further transfers. These results indicate that lens epithelial cells behave similarly to fibroblasts during aging in vitro with respect to the longevity and age of the donor animals.

The question whether the maximum number of CPD is genetically determined or whether it is due to an accumulation of somatic mutations is still open. Both hypotheses are discussed in the literature and both are supported by experimental findings [*Orgel*, 1963; *Holliday* et al., 1977; *Kontermann and Bayreuther*, 1978].

Repair systems, which discern distinct damages of cellular DNA and which are capable of replacing damaged sites have often been investigated, since it seems reasonable to explain age-induced alterations by a decrease in the cellular repair capacity. A close relationship between the increasing life span of a species and the increasing UV excision repair capacity supports this idea [*Hart and Setlow*, 1974]. Therefore the unscheduled DNA synthesis (UDS, repair synthesis) in lens epithelial cells at different CPD levels was investigated by autoradiography. The grains in non-S-phase cell nuclei in

Table II. UV excision repair capacity (UDS) of bovine and rat lens epithelial cells at different CPD levels

Species	CPD	UDS: grains/nucleus[1]
Bovine	10, 23	40–45
Rat	5, 10, 15	16–20
Rat (cat$^+$)	1, 3, 5	18–22

[1] 6 h after UV exposure: 400 erg/mm^2.

Table III. Repair of X-ray-induced strand breakage of bovine lens epithelial cells dependent on donor age

Donor age	CPD	Repair, %[1]
4 years	6	90
7 years	3	85
16 years	3	75

[1] 2 h after X-irradiation: 10 Gy.

contact-inhibited cultures were counted according to *Amlacher* [1974]. The results showed no differences in UDS between early and late passage cells. Differences were found, however, between rat and bovine lens epithelial cells (table II). The data of table II give also the results from cells of a cataract mutant (cat$^+$) [*Gorthy*, 1979] which shows lens opacities soon after birth. Compared to the normal rat strain even in this case no differences in UDS could be found. This means that lens opacities in the mutant cat$^+$ are probably not due to a defect in its UV excision repair system.

Because UV excision repair is only one of the known repair systems [*Laskowski*, 1981], another study was performed. DNA strand breaks induced by X-rays and their repair were measured according to the method of *Ahnström and Erixon* [1973]. No differences dependent on the CPD level could be detected. The capacity to rejoin X-ray-induced DNA strand breaks, however, decreased with increasing donor age (table III).

These experiments demonstrate that the repair capacity remains constant during subcultivation, but they do not exclude the possibility that the repair capacity may decrease in senescent cells of phase III and they cannot

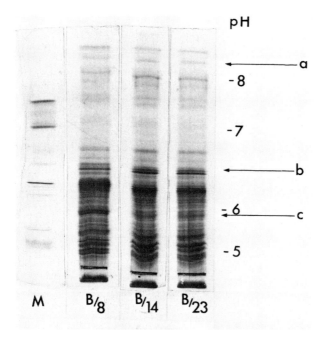

Fig. 3. Protein patterns of diploid rat lens epithelial cell at different CPD levels (8, 14, 23 CPD). M = Marker proteins. Arrows indicate alterations. [From *Rink and Vornhagen*, 1980].

be generalized because only two of several repair systems have been tested so far.

When the protein pattern of cultured lens epithelial cells was investigated during aging in vitro qualitative and quantitative alterations were found [*Rink and Vornhagen*, 1980]. Water-soluble proteins from lens epithelial cells were analyzed by isoelectric focusing in polyacrylamide gels [*Bours*, 1971]. Figure 3 demonstrates three typical alterations which occur with increasing CPD numbers. The band at pH 8.4 (fig. 3, a) diminishes, the triple band at pH 6.4 (fig. 3, b) increases, whereas the band at pH 5.7 (fig. 3, c) disappears completely. The bands in question have been scanned and are shown in figure 4a and b. It is of interest to note that pH 5.7 protein also disappears in the protein pattern of whole rat lenses with increasing age [*Bours*, 1977].

In order to investigate the presence of crystallins, specific antisera against crystallins had been prepared by immunizing young rabbits. Using one-dimensional immunoelectrophoresis it could be demonstrated that the

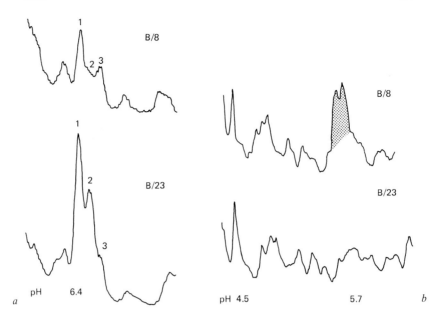

Fig. 4. a Scans of the pH 6.4 region at CPD 8 and 23 according to figure 3. *b* Scans of the pH 5.7 region at CPD 8 and 23 according to figure 3. [From *Rink and Vornhagen*, 1980].

crystallin synthesis was depressed already after a few passages (fig. 5). While in most primary cultures α- and some β-crystallins are detectable, crystallins disappear some passages later. Figure 6 shows a comparable experiment using two-dimensional immunoelectrophoresis [*Laurell*, 1965] of rat lens epithelial cells. The β-crystallin content decreases rapidly from passage 7 to 15 to 22. Similar results could be obtained with the method of indirect immunofluorescence, where fixed cells were incubated with a corresponding antiserum, then washed and incubated with an antirabbit FITC antiserum. Using this method, a few α-crystallin-positive cells could be detected in the center of cell colonies, i.e. in contact-inhibited, nonproliferating cells. This remarkable finding led to experiments with arrested cell cultures, first described by *Dell Orco* [1974]. Mitotic arrest was obtained by lowering the concentration of fetal calf serum to 0.5% over a period of 40–60 days without cell transfer. The protein analyses of such arrested cells are shown in figure 7. While nonarrested cells contain only one weak precipitin line (sample 1), sample 3 with arrested cells shows the presence of several β-lines, one α-line

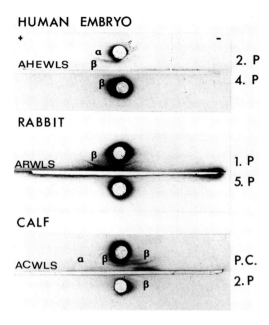

Fig. 5. Immunoelectrophoresis of water-soluble proteins from lens epithelial cells derived from different species against specific antisera. AHEWLS = Antihuman embryonic whole lens serum; ARWLS = antirabbit whole lens serum; ACWLS = anticalf whole lens serum; PC = primary culture; P = passage number. [From *Vornhagen*, 1982].

and a weak γ-line. The presence of γ-crystallin could also be proven by indirect immunofluorescence.

These experiments clearly showed that diploid lens epithelial cells do not lose the capacity to express crystallins during serial subcultivation. This capacity, however, could be detected only in nongrowing mitotically arrested cultures. Obviously the crystallins are expressed in the G_0/G_1 phase of the cell cycle. In well-growing, that means in rapidly proliferating cultures, there is neither time nor need to synthesize these types of proteins (luxury proteins). It should be mentioned that cells from arrested cultures which are inoculated under normal conditions (20% of fetal calf serum) after the time of arrest retain their normal growth behavior, whereas synthesis of crystallins no longer occurs.

Experiments with mitotically arrested cultures [*Vornhagen and Rink*, 1981] have been performed at different CPD levels. Comparable results were

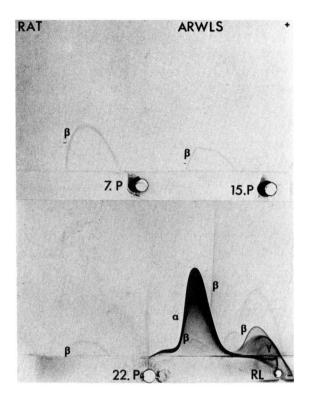

Fig. 6. Two-dimensional Ag/Ab crossed immunoelectrophoresis of water-soluble proteins from rat lens epithelial cells at different passage numbers (7, 15, 22) against antirat whole lens serum (ARWLS). RL = Reference pattern of whole rat lens proteins. [From *Vornhagen*, 1982].

obtained with one remarkable exception. Bovine lens epithelial cells at CPD 68, just before entering the degenerating phase (phase III), do not show any expression of crystallins even after a long time of mitotic arrest. Neither α- nor β- nor γ-crystallins could be detected in this case. Figure 8 shows the immunofluorescence of mitotically arrested bovine lens epithelial cells after 14 and 68 CPD, respectively.

Cells arrested after 14 CPD exhibit positive fluorescence against anti-β-crystallin antiserum, whereas the senescent cells (CPD 68) immediately before entering the phase of degeneration have definitely lost the capacity to synthesize any crystallins. Should this phenomenon also occur in vivo it could explain why in old age, when lens epithelial cells lose their proliferating

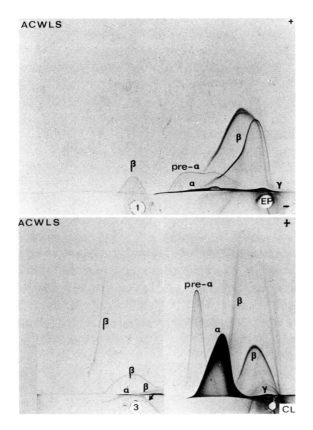

Fig. 7. Two-dimensional Ag/Ab crossed immunoelectrophoresis of water-soluble proteins from calf lens epithelial cells before (1) and after (3) mitotic arrest for 53 days against anti-calf whole lens serum (ACWLS). EP = Proteins of noncultivated epithelia; CL = proteins from whole calf lens. [From *Vornhagen*, 1982].

potential, the synthesis of crystallins disappears. Since normal lens transparency depends on the presence and the state of crystallins, it may be assumed that cells with a defective protein synthesis are likely to disturb the maintenance of lens transparency.

Discussion

The mechanisms involved in the aging behavior of diploid mammalian cells, as described by *Hayflick* [1965], are still unclear. Nevertheless, these

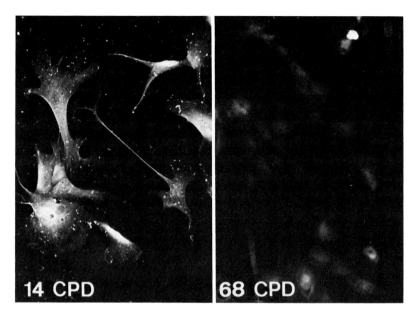

Fig. 8. Immunofluorescence of β-crystallins in calf lens epithelial cells arrested at different CPD levels (14, 68). × 110. [From *Vornhagen*, 1982].

findings stimulated a lot of experiments on the cellular and biochemical levels and increased tremendously our present knowledge on cellular aging [*Cristofalo*, 1972; *Goldstein and Moermann*, 1976; *Kontermann and Bayreuther*, 1978].

A suitable model for such studies are experiments with the population of lens epithelial cells [*Courtois* et al., 1978; *Rink and Vornhagen*, 1979]. They are the only proliferative cells of this organ and they synthesize organ-specific proteins, the crystallins. With respect to aging processes the lens is a highly interesting organ because of the age-related loss of transparency. Our knowledge on the relations between the behavior of lens epithelial cells and the process of lens opacification is rare, but not negligible. During the investigation of human cataractous lenses undifferentiated epithelial cells have been found in the posterior region of the lens [*François* et al., 1978]. Morphological investigations in experimentally induced radiation cataracts (a model often used for studies on senile cataracts) [*Bergeder* et al., 1982] show some severe disturbances in the proliferation and differentiation behavior of lens epithelial cells [*von Sallmann*, 1959; *Worgul and Rothstein*, 1975] long before an opacification becomes evident. Many other findings have been reported on the alteration of the crystallin composition during aging in situ [*Hoenders and Bloemendal*, 1981], part of which cannot be reduced to postsynthetic protein modifications. A special aspect is the decrease of the γ-crystallin content in the outer cortex of old lenses, where a high content of the protein is expected since the newly differentiated cells which constitute the outer cortex should synthesize γ-crystallin. All these findings result from studies on whole lenses; they support the idea that alterations of lens epithelial cells as observed in vitro are involved in the alterations which occur in vivo during aging.

The results presented here demonstrate that:
(1) The proliferative capacity of lens epithelial cells is limited.
(2) The proliferative capacity decreases with increasing donor age.
(3) Lens epithelial cells derived from long-living species show higher CPD numbers than those from short-living ones.
(4) Lens epithelial cells of the rat mutant cat$^+$ which exhibits an inborn lens opacification performs only 3–5 CPD compared to the 32–36 CPD of the epithelial cells from normal rat strain.
(5) There is neither a decrease in UV excision repair capacity nor in the capacity to rejoin X-ray-induced strand breaks during serial subcultivation.
(6) The capacity to rejoin X-ray-induced strand breaks decreases with donor age.
(7) The pattern of water-soluble proteins changes during serial subcultivation.
(8) The disappearance of a protein band at pH 5.7 also occurs in whole lenses of older age.
(9) Under distinct culture conditions (mitotically arrested culture) diploid lens epithelial cells have the ability to express crystallins.
(10) Cells shortly before entering phase III have definitely lost the capacity to express crystallins.

The results indicate that investigations on cellular aging in vitro are well suited to expand our knowledge on age-related alterations occurring in vivo. Particularly in age-related alterations of the lens (cataracta senilis) the aging of lens epithelial cells seems to play an essential role.

Acknowledgements

This work was supported by the Deutsche Forschungsgemeinschaft (grant Ri 285/5-6) and was performed within the framework of a concerted action of the European Community

'EURAGE'. Some of the experiments were performed by R. *Vornhagen* and R. *Zblewski* as part of their theses. The technical assistance of Mrs. A. *Homrighausen* is gratefully acknowledged.

References

Ahnström, G.; Erixon, K.: Radiation-induced strand breakage in DNA from mammalian cells. Int. J. Radiat. Biol. *23:* 285–289 (1973).

Amlacher, E.: Autoradiographie in Histologie und Cytologie (Fischer, Stuttgart 1974).

Bayreuther, K.: Die genetische Regulation des zellulären, organischen und organismischen Alterns. Verh. dt. Ges. Path. *59:* 101–118 (1975).

Bayreuther, K.: Zellbiologie und Zellbiochemie der nicht teilungsfähigen Fibroblasten. DFG-Kolloquium «Biologie des Alterns», 1982.

Bergeder, H.-D.; Rink, H.; Hockwin, O.: Die Röntgenkatarakt. Ein Modell zur Untersuchung der Kataraktogenesis. Klin. Mbl. Augenheilk. *180:* 13–16 (1982).

Bours, J.: Isoelectric focusing of lens crystallins in thin layer polyacrylamide gels. A method for the detection of soluble proteins in eye lens extracts. J. Chromatogr. *60:* 225–233 (1971).

Bours, J.: The crystallins of the aging lens from five species. Studies by various methods of thin layer isoelectric focusing; in Radola, Graesslin, Electrofocusing and isotachophoresis, pp. 303–312 (de Gruyter, Berlin 1977).

Carrel, A.: The permanent life of tissue outside of the organism. J. exp. Med. *15:* 516 (1912).

Courtois, Y.; Counis, M.F.; Laurent, M.; Simmoneau, L.; Tréton, J.: In vitro cultivation of bovine, chicken and human epithelial lens cells in aging studies. Interdiscipl. Topics Geront., vol. 12, 2–12 (Karger, Basel 1978).

Cristofalo, V.J.: Animal cell cultures as a model for the study of aging. Adv. Gerontol. Res. *4:* 45–79 (1972).

Dell Orco, R.T.: Maintenance of human diploid fibroblasts as arrested populations. Fed. Proc. Fed. Am. Soc. exp. Biol. *33:* 1969–1972 (1974).

François, J.; Victoria-Troncoso, V.; Cansa, K.; Victoria-Ihler, A.: Culture of lens epithelium in senile cataracts. Interdiscipl. Topics Geront., vol. 12, pp. 34–40 (Karger, Basel 1978).

Goldstein, S.; Moermann, E.J.: Defective proteins in normal and abnormal fibroblasts during aging in vitro. Interdiscipl. Topics Geront., vol. 10, pp. 24–43 (Karger, Basel 1976).

Gorthy, W.C.: Development of ocular lens abnormalities in rats with X-ray-induced cataract mutants. Investve Ophthal. vis. Sci. *18:* 939–946 (1979).

Hart, R.W.; Setlow, R.B.: Correlation between DNA excision repair and life span in a number of mammalian species. Proc. natn. Acad. Sci. USA *71:* 2169–2173 (1974).

Hayflick, L.: The limited in-vitro life time of human diploid cell strains. Expl Cell Res. *37:* 614–636 (1965).

Hayflick, L.; Moorhead, P.S.: The serial cultivation of human diploid fibroblasts. Expl Cell Res. *25:* 585 (1961).

Hockwin, O.: Age changes of lens metabolism; in Bredt, Rohen, Aging and development, vol. I, pp. 95–129 (Schattauer, Stuttgart 1971).

Hoenders, H.J.; Bloemendal, H.: Aging of lens proteins; in Bloemendal, Molecular and cellular biology of the eye lens, pp. 279–326 (Wiley, New York 1981).

Holliday, R.; Huschtscha, L.I.; Tarrent, G.M.; Kirkwood, T.B.L.: Testing the commitment theory of cellular aging. Science *198:* 366–372 (1977).

Kontermann, K.; Bayreuther, K.: Experimental evidence for a unifying concept of the molecular mechanism of the cellular aging and the cellular neoplastic transformation. Akt. Gerontol. *8:* 411–418 (1978).

Laskowski, W.: Biologische Strahlenwirkung und ihre Reparatur (de Gruyter, Berlin 1981).

Laurell, C.B.: Antigen/antibody crossed electrophoresis. Analyt. Biochem. *10:* 358–361 (1965).

Orgel, L.E.: The maintenance of the accuracy of protein synthesis and its relevance to aging. Proc. natn. Acad. Sci. USA *49:* 517–521 (1963).

Papaconstantinou, J.: Molecular aspects of lens cell differentiation. Science *156:* 338–346 (1967).

Rink, H.; Vornhagen, R.: Crystallins of lens epithelial cells during aging and differentiation. Ophthalmic Res. *11:* 355–359 (1979).

Rink, H.; Vornhagen, R.: Crystallins of rat and bovine lens epithelial cells during aging; in Regnault, Hockwin, Courtois, Aging of the lens, pp. 37–51 (Elsevier, North-Holland, Amsterdam 1980).

Rink, H.; Vornhagen, R.; Koch, H.-R.: Rat lens epithelial cells in vitro. I. Observations on aging, differentiation and culture alterations. In Vitro *16:* 13–19 (1980).

Sallmann, L. von: Early lenticular lesions resulting from ionizing radiation. Trans. Am. Acad. Ophthal. Oto-lar. *63:* 439–488 (1959).

Vornhagen, R.: Immunologisch-biochemische Untersuchungen zur Dedifferenzierung und Differenzierung kultivierter Linsenepithelzellen verschiedener Spezies und verschiedenen Alters. Dissertationsschr. Bonn (1982).

Vornhagen, R.; Rink, H.: The expression of crystallins in early and late passages of calf lens epithelial cells grown under different culture conditions. Eur. J. Cell Biol. *24:* 25 (1981).

Wanko, T.; Gavin, M.A.: Electronmicroscopy of lens fiber. J. biophys. biochem. Cytol. *6:* 97–102 (1959).

Worgul, B.V.; Rothstein, H.: Radiation cataract and mitosis. Ophthalmic Res. *7:* 21–32 (1975).

Prof. H. Rink, Institute of Radiobiology, University of Bonn,
Sigmund-Freud-Strasse 15, D-5300 Bonn (FRG)

III. Immune Senescence

The Cellular Basis of Immune Senescence

Marc E. Weksler, Gregory W. Siskind

Department of Medicine, Cornell University Medical College, New York, N.Y., USA

Introduction

The involution of the thymus after sexual maturity plays a crucial role in immune senescence. This view of the immunobiology of aging has guided our research studies. Our review places in perspective a body of work concerning immune senescence performed in the authors' laboratory during the past 6 years. It is likely that in the future genetic or cellular engineering or immunotherapeutic approaches to immune senescence will be possible. We believe such approaches must be undertaken with the caution as there is a real possibility that the immune deficiency associated with aging may be an adaptive mechanism to protect against the autoimmune reactions that develop with increasing age. It is, therefore, essential to extend our understanding of the age-related changes in the immune system so that immunotherapeutic interventions, as they become available, can be applied with intelligence.

Thymic Involution and Aging

The involution of the thymus after puberty was recognized 30 years before the immunologic function of the organ was discovered. The mass of the human thymus is well maintained until 15 years of age and then rapidly decreases [*Boyd*, 1932]. By the age of 45 or 50, the cellular mass of the human thymus is only 10% of its maximal mass.

Differentiation of Lymphocytes

Precursors of T lymphocytes arise in the bone marrow and migrate to the thymus where their differentiation continues. Both the microenvironment of the thymus and hormones produced by the thymus are important for the differentiation of lymphocytes. The capacity of the human thymus to facilitate the differentiation of T lymphocytes declines with age [*Singh and Singh*, 1979]. This, we found, was associated with the appearance of increased numbers of immature T lymphocytes in the peripheral blood [*Moody* et al., 1981].

The age-associated decline in the capacity of the thymus to mediate the differentiation of immature T lymphocytes has also been demonstrated by transplanting thymus glands from mice of different ages into young syngeneic thymectomized recipients [*Hirokawa and Makinodan*, 1975]. Thymus glands from animals over 3 months of age showed a progressive age-dependent loss in their capacity to mediate the differentiation of T lymphocyte precursors.

The influence of the thymus on lymphocyte differentiation had been thought to be limited to T lymphocytes. Recently, however, we have found that the thymus also plays a role in the differentiation of immature B lymphocytes [*Szewczuk* et al., 1980]. It was shown that the capacity of thymocytes to mediate the differentiation of immature B lymphocytes declined with age of the thymocyte donor.

Endocrine Function of the Thymus

During the past 10 years, a number of thymic hormones, e.g. thymopoietin, thymosin alpha-1, and 'facteur thymique sérique', have been insolated and purified. All these hormones decline with age. The level of thymopoietin in human serum is maintained between birth and 30 years of age but thereafter declines and becomes undetectable after age 60 [*Lewis* et al., 1978]. The concentration of 'facteur thymique sérique' in human serum begins to fall after the age of 20 and becomes undetectable after the age of 50 [*Bach* et al., 1972]. The concentration of thymosin alpha-1 in human serum appears to decline even earlier [*Goldstein*, personal commun., 1983]. Thus, the decline in thymic mass is followed by a progressive loss in thymic differentiation of T lymphocytes and by a decline in the serum concentration of thymic humoral factors.

Peripheral T Lymphocyte Populations

Although the total number of T and B lymphocytes in the peripheral blood mass does not change significantly with age [*Gupta and Good*, 1979], the distribution of T lymphocyte subpopulations is altered. We have found that the number of immature T cells, and T lymphocytes identified by the

OKT4 monoclonal antibody increases with age, while the number of T lymphocytes identified by the OKT8 monoclonal antibody decreases with age [*Moody* et al., 1981].

Senescence of the Immune Response

Humoral Immunity

The total serum concentration of immunoglobulins does not change significantly with age. However, the distribution of immunoglobulin classes does change. The serum concentrations of IgA and IgG increase and IgM decreases with age. It has been shown that there is a decrease in the concentration of isoagglutinins and natural antibody to sheep erythrocytes (SRBC) and a decrease in the antibody response to Japanese B encephalitis virus, parainfluenza virus, pneumococcal polysaccharide, salmonella flagellin and tetanus toxoid is depressed in old humans. In contrast to the decline with age in antibodies to foreign antigens, the incidence of autoantibodies increases markedly with age [*Hallgren* et al., 1973]. Autoimmune disease, however, is not increased in the elderly. We found that although old animals have a higher incidence of autoantibody production and more autoantibody producing cells, they are no more susceptible to autoimmune disease than young animals [*Goidl* et al., 1981]. Autoantibodies of particular interest are auto-anti-idiotype antibodies involved in the normal regulation of the immune response. Their increased production could be one of the mechanisms responsible for the decreased immune response associated with aging.

Another manifestation of altered humoral immunity associated with age is the increasing prevalence of benign monoclonal gammopathies. The increased incidence of monoclonal immunoglobulins in aged humans and animals probably is another manifestation of the loss of thymic function since neonatal thymectomy lowers the age at which monoclonal gammopathies appear and increases the percentage of animals which ultimately develop them [*Radl* et al., 1980].

Cell-Mediated Immunity

Cell-mediated immunity depends on the functional integrity of T lymphocytes. Delayed hypersensitivity reactions and graft rejection, two classic manifestations of cell-mediated immunity in vivo, decline with age. Lymphocytes from old mice are less capable of eliciting a graft-vs-host reaction or of rejecting tumor or skin grafts than are lymphocytes from young mice.

However, incubation of lymphocytes from old animals with thymic hormone restored their capacity to manifest a vigorous in vitro graft-vs-host reaction [*Friedman* et al., 1974]. The increased susceptibility of old animals to many infections has been related to a failure of T cell immunity. For example, old mice infected with *Listeria monocytogenes*, develop only one-thousandth the level of T cell immunity of that in young mice [*Patel*, 1981].

A syngeneic graft-vs-host reaction has been demonstrated following the transfer of lymphocytes from old animals to syngeneic recipients. This reaction does not occur when lymphocytes from young animals are transferred [*Gozes* et al., 1978]. These observations suggest that increased numbers of 'autoimmune' T lymphocytes as well as B lymphocytes develop with age.

Several in vitro correlates of T lymphocyte-mediated immunity decline with age. These include: (1) the proliferative response of T lymphocytes to the mitogens; (2) the proliferative response of T lymphocytes to allogeneic or autologous non-T cells, and (3) the generation of cytotoxic T lymphocytes [*Weksler and Hutteroth*, 1974; *Moody* et al., 1981].

Cellular Basis of Immune Senescence

We have studied the cellular basis of the age-associated defects in the immune system both in vitro and in vivo using cell transfer methods. Both methods permit the intrinsic function of lymphocytes to be evaluated in a controlled environment in the absence of influences from the other physiological changes that accompany aging which might indirectly compromise immune function. Data obtained by both experimental approaches are consistent with the hypothesis that the predominant age-associated defect in immune function is in the T lymphocyte population.

We have studied in detail the cellular basis of the impaired proliferation of T lymphocytes from old humans when cultured with mitogens. The defect is intrinsic to the T lymphocyte populations and not due to influences of other cell populations, as highly purified T lymphocytes from old humans express the proliferative defect. There is no demonstrable difference in the number or affinity of receptors for phytohemagglutinin (PHA) on T lymphocytes with age. We found that, despite having normal number of T lymphocytes, healthy old persons have only one-half the number of mitogen-responsive T lymphocytes found in the peripheral blood of young subjects. This conclusion is based upon three independent experimental approaches: (1) limiting dilution analysis of the number of PHA-responsive lymphocytes; (2) determination of the number of T lymphocytes activated in culture with mitogens by their susceptibility to infection with the vesicular stomatitis virus, and (3) the incorporation of tritiated thymidine by lymphocytes cultured with mitogen in the presence of colchicine [*Inkeles* et al., 1977].

We found that not only is the number of mitogen-responsive T cells reduced in old humans, but the capacity of these lymphocytes to divide sequentially in culture is also impaired [*Hefton* et al., 1980]. This was established by quantitating the number of replicative cycles completed by lymphocytes from old and young donors in culture with PHA. We found that while the number of T lymphocytes dividing for the first time was comparable in cultures from old and young donors the number of T lymphocytes from old donors dividing for a second or third time was only one-half and one-quarter, respectively, of that observed in cultures from young donors. Since there is no difference in the length of the cell cycle in young or old donors, these findings indicate a defect in the capacity of lymphocytes from old subjects to divide. Similar defects in the proliferative capacity of fibroblasts [*Martin* et al., 1970] and arterial smooth muscle cells [*Bierman*, 1978] from old humans have been reported.

The molecular basis of the proliferative defect observed in lymphocytes from old humans has also been investigated in our laboratory. Recent studies have indicated that a T lymphocyte product, interleukin 2 (IL-2), is necessary for T cell proliferation. We found that lymphocytes from old humans produce only half as much IL-2 as do lymphocytes from young donors [*Gillis* et al., 1981]. In addition, lymphocytes from old humans are less responsive to IL-2. This appears to be due, at least in part, to a reduced capacity of lymphocytes from old donors to bind IL-2. This may be due to a decrease in the number and/or the affinity of IL-2 receptors.

Recently, techniques have been developed to study antibody production by human lymphocytes in culture. This has permitted us to examine the cellular basis of the impaired humoral immune response of elderly humans [*Kim* et al., 1981]. Formalinized staphylococci stimulate both T-independent and T-dependent polyclonal antibody secretion by lymphocytes in culture. This assay revealed that less antibody is produced by unfractionated human lymphocytes from old as compared to young donors. In contrast, there was no significant difference in antibody production by B lymphocytes from old and young donors. These results suggest that the age-associated defect in antibody production by lymphocytes from old humans in vitro is expressed mainly by the T lymphocyte population. T lymphocytes from young humans augment antibody production while T lymphocytes from the elderly suppress it. In addition, we found that staphylococci or antihuman immunoglobulin antibody stimulate the same amount of thymidine incorporation by B lymphocytes from young and old subjects, again suggesting that B lymphocytes from the aged are relatively unimpaired [Kim et al., 1981].

In addition to B and T lymphocytes, macrophages play a critical role in the immune response. They appear to be involved in antigen presentation as well as in the secretion of factors which stimulate lymphocytes. Macrophages incubated with lipopolysaccharide (LPS) release a T lymphocyte replacing factor (TRF) which permits the in vitro secretion of antibody to SRBC by purified murine B lymphocytes. There is no difference in TRF production by macrophages from old or young humans cultured with LPS [Kim et al., 1981]. Thus, macrophages from elderly humans are not impaired in function, at least with respect to the production of TRF. In summary, the cellular basis of the impaired immune response of elderly humans appears to rest with T lymphocyte population which expresses two defects: (1) fewer responsive lymphocytes, and (2) impaired proliferative capacity of the responsive lymphocytes. Within the limits of available assays, the functions of macrophages and B lymphocytes are appear to be relatively well maintained with age.

Mice also show an age-related decline in humoral and cell-mediated immunity. It has been known for some time that the antibody response to the T-dependent antigens is markedly impaired with age. We have characterized this defect in humoral immunity associated with aging using chemically defined hapten protein conjugates. Studies using the T-dependent antigen, dinitrophenylated bovine gamma-globulin (DNP-BGG), revealed that in addition to the age-related decline in total antibody production, there was a preferential loss of high affinity and of IgG antibodies [Goidl et al., 1976]. The preferential loss of high affinity antibodies has implications both for the health of old animals and for the mechanism underlying immune senescence. High affinity antibody probably affords more effective protection against infection than does low affinity antibody. The increased susceptibility of old animals to infection may reflect, at least in part, the loss of high affinity antibody production. Surface-bound antibody on lymphocytes is the receptor for antigen. Consequently, the binding of antigen by the lymphocyte, one of the signals required for its activation depends on the affinity of surface-bound antibody. If the affinity of surface-bound antibody is low, higher concentrations of antigen are required to activate the lymphocyte. The loss of high affinity receptors for antigen on lymphocytes probably explains the previously reported observation that higher doses of antigen are necessary to elicit a maximum immune response in old as compared with young animals

[*Makinodan and Adler*, 1975]. A clinical consequence suggested by these findings is that the multiplication of an infectious agent would continue longer in the elderly before an immune response was initiated and host immune defense mechanisms developed.

Self-tolerance is the immune mechanism which prevents autoimmunity. We showed that increased doses of antigen are required to induce immunologic unresponsiveness in old as compared to young animals [*Dobken* et al., 1980; *DeKruyff* et al., 1980b]. In part, this might be a reflection of the decrease in high affinity antibody-producing cells in old mice, as such cells are known to be more readily rendered tolerant. It is possible that low concentrations of autologous antigens, which are sufficient to maintain self-tolerance in young animals, do not maintain self-tolerance in old animals. Apparently with constant stimulation by self-antigens and impaired self-tolerance, old mice develop increased numbers of autoreactive B cells and increased production of autoantibodies [*Goidl* et al., 1981].

Among the autoantibodies produced in greater amount by old mice are auto-anti-idiotype antibodies [*Goidl* et al., 1980]. Anti-idiotype antibodies are specific for the antigen combining site of another antibody molecule. In the systems we have studied, such autoantibodies inhibit or 'down regulate' the immune response. The enhanced auto-anti-idiotypic antibody response in aged mice therefore not only results from but contributes to immune senescence. In addition, our studies revealed that the repertoire of idiotypes changes with age. That is, the DNP-specific idiotypes generated during the immune response of young and old mice to a haptenic determinant are different. Thus, both quantitative and qualitative (isotype, affinity and idiotype repertoire) changes in the immune response occur with age.

We have extensively investigated the cellular basis for the age-associated defects in the immune response of experimental animals by using in vitro cell culture methods. A primary specific plaque-forming cell (PFC) response to either a T-dependent or a T-independent antigen can be generated in cultures of murine spleen cells. We found that the in vitro PFC response to the T-dependent antigen, SRBC, was markedly impaired with cells from old as compared to young animals [*DeKruyff* et al., 1980a]. The anti-SRBC response was compromised by 12 months of age and by 18 months of age, the response was less than 5% of the maximal response observed in mice between the ages of 3 and 6 months. We found that deficiency of helper T lymphocytes, excessive suppressor T cell activity and an intrinsic defect in B cell function contribute to the defective anti-SRBC response of cells from old mice. The intrinsic defect in B cells was evidenced by the fact that the anti-

DNP PFC response of spleen cells from aged mice to the T-independent antigen DNP-polyacrylamide (DNP-PAA) beads was impaired even after the spleen cell population was depleted of T cells. This implies that the reduced PFC response cannot be due solely to the increased suppressor activity found in the T lymphocyte population from old animals but must be due, at least in part, to a defect in the function of the peripheral B lymphocyte population. This may reflect a failure of thymocyte-dependent B lymphocyte maturation [*DeKruyff* et al., 1980a].

Cell transfer studies in inbred mice have provided additional insight into the immune deficiency of aging. In such studies, the function of lymphocytes from old and young animals are compared after their transfer into lethally irradiated syngeneic young recipients. This technique permits a direct assessment of the in vivo function of lymphocytes from old and young mice in the absence of influence by other age-associated physiological or pathological changes that may exist in old animals.

Classic studies by *Makinodan and Adler* [1975], established that 90% of the age-associated defects in the antibody response of mice to SRBC are due to intrinsic defects to the peripheral lymphoid cell population of the aged animals. Only 10% of the defective response can be attributed to the 'internal environment' of the old host. Using similar cell transfer techniques, we found that the preferential loss of high affinity and IgG antibody production by old animals is also the result of intrinsic changes in the peripheral lymphoid cells. In mixed cell transfer studies in which irradiated mice were reconstituted with a mixture of cells from young and old donors, we were able to demonstrate a marked increase in suppressor cell activity in aged mice [*Goidl* et al., 1976].

We observed that the suppressor cells present in aged mice did not preferentially affect high affinity antibody production. This suggests that increased suppressor activity, while probably contributing to the immune defect of aging, does not totally account for it. The in vivo studies corroborate those we obtained in vitro where deletion of suppressor T cells from the spleen cell population of aged mice increased their response to a T-independent antigen but did not restore the response to the level achieved by cells from young mice [*DeKruyff* et al., 1980a].

The fact that young mice reconstituted with spleen cells from old donors manifest the impairments in high affinity antibody production and IgG PFC response which are characteristic of old animals permitted us to define precisely the mechanism of the defective response. The high affinity and IgG antibody responses of irradiated recipients of spleen cells from old donors are restored if thymocytes from young donors are transferred together with the spleen cells from old donors [*Goidl* et al., 1976]. Furthermore, residence of spleen cells from old donors for 8 weeks in young, irradiated recipients possessing an intact thymus gland reversed the age-associated defects in high affinity and IgG antibody production. No reversal of the age-associated immune defects was seen in thymectomized recipients of spleen cells from old donors. Finally, we found that these age-associated defects in the spleen cell population from old mice of old lymphocytes can be reversed by incubating old lymphocytes with the thymic hormone prior to cell transfer [*Weksler* et al., 1978]. The importance of thymic involution in immune senescence was supported by our finding that adult thymectomy accelerates and injection of thymic hormone delays the appearance of age-associated immune defects [*Weksler* et al., 1978].

The cellular basis for the increased auto-anti-idiotype antibody response of aged mice was also studied by cell transfer methods. We found that the increased production of auto-anti-idiotype antibody was transferred by spleen cells but not by bone marrow cells from old donors. Mixed cell transfer experiments showed that peripheral T lymphocytes are responsible for the increased auto-anti-idiotype antibody response of old mice [*Goidl* et al., 1983].

The fact that two immune mechanisms, suppressor cell activity and auto-anti-idiotype antibody production, which down regulate the immune system, are increased with age suggests that a failure of these regulatory mechanisms is not the cause of autoantibody production in the aged. In fact, the age-related increase in the activity of mechanisms involved in down regulating the immune response suggests that they might actually be activated as protective mechanisms in an effort to prevent autoantibody production. According to this view, the decreased immune response to foreign antigens that accompanies aging may be an undesirable side effect of homeostatic mechanisms which protect the organism against the consequences of autoantibody production.

Conclusion

We have demonstrated a number of factors which contribute to the decreased immune response seen in aged humans and experimental animals. These include: (1) decreased helper T lymphocyte activity; (2) increased suppressor T lymphocyte activity; (3) increased auto-anti-idiotype antibody production; (4) a decreased number of T lymphocytes responsive to mitogen; (5) a decreased proliferative capacity of T lymphocytes to mitogens, and (6) a decreased production of IL-2 by T lymphocytes and a reduced response of T lymphocytes to IL-2.

These data support the primacy of thymic involution in immune senescence. Experiments demonstrating a reversal of certain age-associated defects in immune function following transfer of thymocytes from young donors to old recipients, transfer of old lymphocytes to young donors with intact thymus glands, or treatment of old mice with thymic hormone, support this conclusion. These findings offer potential therapeutic strategies for reversing immune senescence. However, it is important to bear in mind that the progressive programmed decline in thymic function may well be an evolutionarily developed protective mechanism against overwhelming autoimmune damage.

References

Bach, J.F.; Papiernick, M.; Levasseur, P.; Dardenne, M.; Baros, A.; LeBrigand, H.: Evidence for a serum-factor secreted by the human thymus. Lancet *ii:* 1056 (1972).

Bierman, E.L.: The effect of donor age on the in vitro life span of cultured human arterial smooth-muscle cells. In Vitro *14:* 951 (1978).

Boyd, E.: The weight of the thymus gland in health and in disease. Am. J. Dis. Child. *43:* 116 (1932).

DeKruyff, R.H.; Kim, Y.T.; Siskind, G.W.; Weksler, M.E.: Age related changes in the in vitro immune response: increased auppressor activity in immature and aged mice. J. Immun. *125:* 142 (1980a).

DeKruyff, R.H.; Rinnooy-Kan, E.A.; Weksler, M.E.; Siskind, G.W.: Effect of aging on T-cell tolerance induction. Cell. Immunol. *56:* 58 (1980b).

Dobken, J.; Weksler, M.E.; Siskind, G.W.: Effect of age on ease of B-cell tolerance induction. Cell. Immunol. *55:* 66 (1980).

Friedman, D.; Keiser, V.; Globerson, A.: Reactivation of immunocompetence in spleen cells of aged mice. Nature, Lond. *251:* 545 (1974).

Gillis, S.; Kozak, R.W.; Durante, M.; Weksler, M.E.: Immunological studies of aging. Decreased production of and response to T-cell growth factor by lymphocytes from aged humans. J. clin. Invest. *67:* 937 (1981).

Goidl, E.A.; Choy, J.W.; Gibbons, J.J.; Weksler, M.E.; Thorbecke, G.J.; Siskind, G.W.: Production of auto-anti-idiotype antibody during the normal immune response. J. exp. Med. *157:* 1635 (1983).

Goidl, E.A.; Innes, J.B.; Weksler, M.E.: Immunological studies of aging. II. Loss of IgG and high avidity plaque-forming cells and increased suppressor cell activity in aging mice. J. exp. Med. *1445:* 1037 (1976).

Goidl, E.A.; Michelis, M.A.; Siskind, G.W.; Weksler, M.E.: Effect of age on the induction of autoantibodies. Clin. exp. Immunol. *44:* 24 (1981).

Goidl, E.A.; Thorbecke, G.J.; Weksler, M.E.; Siskind, G.W.: Production of auto-anti-idiotype antibody during the normal immune response. Changes in auto-anti-idiotypic antibody response and the idiotype repertoire associated with aging. Proc. natn. Acad. Sci. USA *77:* 6788 (1980).

Gozes, Y.; Umiel, T.; Ashes, M.; Trainin, N.: Syngeneic GvH induced in popliteal lymph nodes by spleen cells of old C57BL/6 mice. J. Immun. *121:* 2199 (1978).

Gupta, S.; Good, R.A.: Subpopulation of human T lymphocytes. X. Alterations in T, B, third population cells and T cell with receptors for immunoglobulin M or G in aging humans. J. Immun. *122:* 1214 (1979).

Hallgren, H.M.; Buckley, C.E.; Gilbertsen, V.A.; Yunis, E.J.: Lymphocyte phytohemagglutinin responsiveness, immunoglobulins and autoantibodies in aging humans. J. Immun. *111:* 1101 (1973).

Hefton, J.M.; Darlington, G.; Casazza, B.A.; Weksler, M.E.: Immunological studies of aging. V. Impaired proliferation of PHA responsive human lymphocytes in culture. J. Immun. *125:* 1007 (1980).

Hirokawa, K.; Makinodan, T.: Thymic involution. Effect on T differentiation. J. Immun. *114:* 1659 (1975).

Inkeles, B.; Innes, J.B.; Kuntz, M.M.; Kadish, A.S.; Weksler, M.E.: Immunological studies of aging. III. Cytokinetic basis for the impaired response of lymphocytes from aged humans to plant lectins. J. exp. Med. *145:* 1176 (1977).

Kim, Y.T.; Siskind, G.W.; Weksler, M.E.: Cellular basis of the impaired immune response of elderly humans; in Fauci, Ballieux, Human B cell functions: activation and immuno-regulation (Raven Press, New York 1981).

Lewis, V.M.; Twomey, J.J.; Bealmear, P.; Goldstein, G.; Good, R.A.: Age, thymic involution and circulating thymic hormone activity. J. clin. Endocr. Metab. *48:* 145 (1978).

Makinodan, T.; Adler, W.H.: Effects of aging on the differentiation and proliferative potentials of cells of the immune system. Fed. Proc. *34:* 153 (1975).

Martin, G.M.; Sprague, C.A.; Epstein, C.J.: Replicative life span of cultivated human cells. Effects of donor's age, tissue and genotype. Lab. Invest. *23:* 86 (1970).

Moody, C.E.; Innes, J.B.; Staiano-Coico, L.; Incefy, G.S.; Thaler, H.T.; Weksler, M.E.: Lymphocyte transformation induced by autologous cells. Immunology *44:* 431 (1981).

Patel, P.S.: Aging and antimicrobial immunity. J. exp. Med. *154:* 821 (1981).

Radl, J.; DeGlopper, E.; Van den Berg, P.; Van Zwieten, M.J.: Idiopathic paraproteinemia. III. Increased frequency of paraproteinemia in thymectomized aging C57BL/KaLwRij and CBA/BrRij mice. J. Immun. *125:* 31 (1980).

Singh, V.; Singh, A.K.: Age-related changes in human thymus. Clin. exp. Immunol. *37:* 507 (1979).

Szewczuk, M.R.; DeKruyff, R.H.; Goidl, E.A.; Weksler, M.E.; Siskind, G.W.: Ontogeny of B lymphocyte function. VII. Failure of thymus cells from aged donors to induce the functional maturation of B lymphocytes from immature donors. Eur. J. Immunol. *10:* 918 (1980).

Weksler, M.E.; Hutteroth, T.H.: Impaired lymphocyte function in aged animals. J. clin. Invest. *53:* 99 (1974).

Weksler, M.E.; Innes, J.B.; Goldstein, G.: Immunological studies of aging. IV. The contribution of thymic involution to the immune deficiencies of aging mice and reversal with thymopoietin. J. exp. Med. *148:* 996 (1978).

M.E. Weksler, MD, The Department of Medicine, Cornell University Medical College, 1300 York Avenue, New York, NY 10021 (USA)

Cellular Aging, Idiotype Repertoire Changes and Mucosal-Associated Lymphoid System[1]

Myron R. Szewczuk, Andrew W. Wade

Department of Microbiology and Immunology, Queen's University, Kingston, Ont., Canada

Over 50 years ago the first description of age-related changes in the immune system were published by *Friedberger* et al. [1] and *Thomsen and Kettel* [2], who described that serum immunoglobulins to blood group antigens and xenogeneic erythrocytes declined with age. Since then attention has focussed on the immune system in studies of aging for various reasons: (a) The thymus (controlling the maturation of T lymphocytes) is the first organ to undergo changes associated with senescence, involuting at sexual maturity in all mammals [3, 4]; (b) genes of the major histocompatibility complex (controlling immune reactivity) when transferred from a long-lived to a short-lived mouse strain, extend the life span of that short-lived mouse [5, 6]; (c) techniques amenable to the study of the molecular and cellular biological changes occurring with age are easily applied to the cells of the immune system, and (d) with age there is an increased incidence of diseases associated with an impaired immune system [7–9].

Several theories have been extended to explain aging in the immune system, ranging from the Unitary Immunologic Theory of Aging where the programmed decline in the physiological competence of the individual is due to changes manifested exclusively through the immune system [reviewed in 10], to less encompassing theories ascribing the immunologic decline in terms of intrinsic or programmed cellular changes [11; reviewed in 12, 13], or the role of extrinsic and microenvironmental factors in the aging process [14–19; reviewed in 20, 23]. It is likely that the decline in the various physiological

---

[1] This work was supported in part by grants-in-aid of research from the Medical Research Council of Canada, Grant No. MA-7347, the Muscular Dystrophy Association of Canada, and the Gerontology Research Council of Ontario.

functions during senescence is due to some aspects of all of the above theories. At present, however, there is little agreement with regard to their relative contributions; several investigators [20–23] have placed the relative responsibility in the aging process at 10 and 90% for extrinsic and intrinsic influences, respectively.

We understand little of the mechanisms underlying the age-related decline in immune function. Efforts for the most part have been concentrated on the rate and degree of the decline of the major components of the systemic immune system. Furthermore, there is no information on the effect of aging on the immune function at different local sites of the immune system, for example, the mucosal lymphoid system.

Aging and the Immune System

There is considerable evidence to suggest that a relationship exists between aging and a decline in B cell function. Thymic (T)-dependent splenic B cell responses are most affected by age, as manifested by a depressed immune response to a variety of T-dependent antigens [19, 24–26, 69]. In addition, splenic B-cell responses to T-independent antigens have been shown to decrease with age [21, 27]. Explanations for the decline include decrease in T helper activity [11, 24, 26, 28–30], increases in suppressor activity [15, 18, 24, 29–33], or changes in the efficiency of cellular interactions required for the response [30, 34, 35]. Intrinsic defects have also been suggested to account for some of this decline, as well as reducing the proliferative capacity of these cells [11, 18, 24, 27, 31a, 33–49, 69].

On the other hand, others have found that the number of B cells in lymph nodes and spleen do not change appreciably with age [50], but that there is an increase in the number of plasma cells, especially in mice of certain autoimmune-prone strains [7]. Studies on the total number of stem cells from mouse bone marrow have shown that they remain quite constant with age [51] and appear not to lose their lymphohematopoietic activity [52]. While these studies indicate that hematopoietic stem cells do not exhibit intrinsic defects with age, events of maturation and differentiation of these cells should also be considered.

Reductions in various murine T cell functions have been found to occur with age; these include decreases in cytotoxic T lymphocyte (CTL) function [57, 58], delayed-type hypersensitivity (DTH) reactions [59], mixed leukocyte reactions (MLR) [57], protective mediator T cells in *Listeria monocytogenes*

[60] and lactate dehydrogenase virus [61] infections, allogeneic effect induction [62] and in helper T cell activity [24, 26, 29, 35, 48, 63]. Conversely, suppressor T cell activity has been found to increase in these animals [24, 26, 29, 31a, 33, 48, 64, 65]. These findings in the mouse are in direct contrast to that found in the human system where an increase in helper cell [36] and a decrease in suppressor cell [36, 47] function have been reported. In this system, it is not entirely clear how increases in helper cell function, as measured by enhancement of immunoglobulin (Ig) production in mixed populations of 'young' B cells and 'aged' $OKT4^+$ (versus 'young' $OKT4^+$) T cells [36], can coexist with the reported decline in interleukin-2 (IL-2) production and reactivity of these cells [66]. It is suspected that the allogeneic cell mixtures used in these experiments, and in the determinations of suppressor cell activity [36, 47], do not give a true representation of these regulatory cell functions.

Detection of T cell changes within the murine immune system with age has relied on mixing experiments with either whole spleen cell populations [24, 25, 31a, 33, 62] or isolated B and T cells [26, 29, 48, 63] from young and old syngeneic mice, using either in vitro experiments [33, 35, 63], adoptive cell transfer experiments [24, 26, 29, 62] or after growth in diffusion chambers [48, 65]. Specific [24, 26, 29, 31a, 33, 48, 62, 65] or nonspecific [31b, 64] suppressor cells for the most part were detected by anti-Thy-1.2 or anti-Lyt-2 antibody plus complement treatment whereby enhancing the plaque-forming cell (PFC) response of 'old' T and 'young' B cell mixtures.

It is noteworthy that in one study [48], simultaneous measurements were made of helper, suppressor and B cell function in individual mice. A unique method of detecting T helper cell activity was employed in this work and bears mention, since the other studies which measured this cell function could not completely exclude the contribution of suppressor cell activity in their assays. Helper function in this study was measured by the ability of T cells, activated with carrier-anticarrier immune complexes which preferentially induces T helper cells, to restore the PFC response of hapten-primed, T-depleted syngeneic cells after incubation in peritoneal cavity-implanted diffusion chambers. Using such a technique, *Liu* et al. [48] were able to confirm previous findings of defective T cell help in aged mice. This activity, and B cell responses to trinitrophenyl (TNP)-Ficoll declined progressively after 4 months of age while suppressor cell activity increased biphasically with a rapid rise until 15 months of age, and a slow increase thereafter in (C57BL/6xC3H)F1 mice. The decline in immune responsiveness with age was found to preceed the detection of suppressor cells in two studies [24, 33], while in another, the decrease in T cell function preceded the development of autoreactive antibody [75]. Therefore, there is evidence for a relationship between the various functional T cell changes and the development of autoimmunity, but the picture is less clear for the role of increased suppressor cell activity in age-related immune response reductions.

The interleukins also appear to play a critical role in these declining T cell functions with age. Several investigators have noted that murine IL-2 production is reduced with age in response to concanavalin A (ConA) [56, 57, 63], Fc fragments [45], antigen [57], phytohemagglutinin (PHA) [63], or allogeneic stimulation [57]. Active suppression was not responsible for the reduction but instead it was suggested to be the result of a defect in the Lyt-1^+ T cell [45, 56, 57, 63] and/or the synthesis of interleukin-1 (IL-1) by macrophages [56]. The addition of exogenous IL-2 to cell cultures was able to restore MLR [57], CTL [57], ConA [56] and PFC [63] responses to young adult levels, while antigen-induced proliferative responses were not completely recovered by such treatment [57].

Thus, it appears that reduced IL-2 production accounts for much of the age-related decline in the various immune parameters of old mice. Since exogenous IL-2 can recover some of this cell-mediated reactivity, a reduction in IL-2 receptors does not seem likely on these senescent cells. At present, the defect cannot be unequivocally attributed to a primary reduction in IL-1 or in IL-2 synthesis. Regardless, the reduced endogenous production of these lymphokines could account, in part, for reductions in both antibody formation and the mitogenic proliferation of B and T cells. For example, antibody production has been shown to be dependent upon the IL-2-induced proliferation of T helper cells [76, 77]. Mitogen stimulation of T cells is also dependent upon proliferative IL-2 signals since mitogens are thought to simply induce IL-2 receptor expression on these cells [reviewed in 76]. The mitogenic stimulation of B cells now is recognized to require the presence of IL-1 and a T cell-derived B cell growth factor [78].

Results of studies on macrophages have shown that their handling of antigens is not affected with age, both during the induction of the immune response and by ingestion of antigenic aggregates [53–55]. However, recent work by *Chang* et al. [56] has indicated that macrophages may play an important role via a decreased production in IL-1 in the reduced IL-2 production noted in aged mice. Thus, further work in this regard is required to define their precise role in the aging process.

Aging and Idiotype Repertoire Changes

It is now evident that, for some antigens, the receptor/idiotype repertoire of B cells undergoes changes with age [46, 67, 68]. This repertoire change in systemic B cells from aged animals was not found in the stem cells of these animals [18, 68], implying that this change must develop upon maturation and differentiation in the systemic environment. Other studies have demonstrated that it may occur as a consequence of concomitant shifts in the receptor repertoire of T cells, through idiotype-anti-idiotype interactions [18, 58, 64].

Evidence that the systemic B cell repertoire change in aged mice is secondary to the influence of the 'aged' microenvironment, comes from a study which showed that stem cells from old animals, upon transfer to young irradiated carrier-primed recipients, differentiate into B cells with a 'young' repertoire and, conversely, stem cells isolated from young animals and allowed to mature in an old environment expressed the 'aged' B cell repertoire [68]. As a corollary, transfer of spleen cells from an old animal into an irradiated young animal resulted in B cell PFC responses that were elevated for a short period but rapidly reverted to the 'aged' B cell response level [22]. The evidence presented in the first example above, demonstrates that the aged environment alone is sufficient to induce the change from a young to an aged B cell repertoire and that intrinsic defects in this regard are not present within the aged stem cell population. The second example illustrates that this microenvironmental influence induces specific cells to effect the repertoire change, since placement of aged spleen cells in a young environment does not permanently reverse immune responsiveness to 'young' levels. Thus, it appears that the stem cell receptor/repertoires of young and old animals are indistinguishable from one another, but do demonstrate exquisite sensitivity to environmental influences and, subsequently, to the effect of T cell-dependent anti-idiotypic regulation.

These findings raise interesting questions regarding recent studies performed in this area. Since it is evident that idiotype/receptor repertoires change with age: (a) Can the mixtures of 'young' and 'old' cells, used in numerous experiments for classifying B cell deficiencies [69, 71, 75], function efficiently in the generation of an immune response, given that idiotype identity is required for an optimal response [67, 68, 70–73]? (b) Does this influence cause net reductions in the frequency of antigen-reactive cell clones in old animals? Evidence against the first question has been presented by *Szewczuk* [67], while evidence for the second one comes from a study by *Zharhary and*

Klinman [74]. Thus, studies describing changes in the B cell compartment have relied on data generated by techniques that may be open to criticism.

With regard to the first question, *Szewczuk* [67] using adoptive transfers of 'young' and 'old' T and B cells into both young and old irradiated recipients, found that only the combination of 'old' B cells with 'old' T cells would generate an anti-idiotypic antibody response, as detected by hapten (TNP) augmentation of anti-TNP PFC. This indicated that, at least for the generation of an anti-idiotypic response, absolute synergy was required between the responding cells, and, furthermore, provided further evidence for the existence of an altered idiotype repertoire in aged mice.

With regard to the second question, classical aging studies have relied on the detection of immune response parameters (for example, secreted Ig and numbers of specific PFC) where a decline with age was interpreted only in terms of there being either intrinsic defects in the responding cells or regulatory imbalances in the animal. In actual fact, results obtained from testing PFC responses to T-independent antigens in vivo [27] or in vitro [33, 35] could not distinguish between an intrinsic B cell defect or a reduction in the number of antigen-specific B cells with age. To this regard, *Zharhary and Klinman* [74] have examined the B cell response to dinitrophenylated hemocyanin (DNP-HY) at the clonal level in an attempt to differentiate between these two possibilities. They found that DNP-responsive B cells in aged mice yielded antibody forming clones that were normal in both the amount and relative affinity of antibody produced, but with slightly less of a tendency to produce IgG antibody. Furthermore, the frequency of anti-DNP specific cells in the bone marrow of these aged mice was equal to that found in young animals, and most interestingly, this frequency was reduced in the aged spleen. This implied that the decline in immune reactivity found with age was due, at least in part, to a decrease in the number of DNP-responsive cells and not to an intrinsic or regulatory defect per se. Anti-idiotypic regulation was thought to be responsible for this decrease in systemic clonal frequency [64]. Further support for a systemic idiotypic/receptor repertoire shift comes from a recent study performed by *Nishikawa* et al. [116], using the antigen (4-hydroxy-3-nitrophenyl)acetyl (NP). They showed that anti-idiotypic antibody alone is capable of effecting changes in the B cell idiotype repertoire during cell maturation.

At the clonal and stem cell level, *Gorczynski* et al. [58] have examined the status of allogeneic precursor CTLs (pCTL) and mature CTLs in aging C57BL/6J, CBA/J, C3H/HeJ, and C3HB6 F1 mice. Using limiting dilution analysis of splenic pCTL in mixed leukocyte cultures, it was found that the frequency of clones for different antigenic determinants changed with age. Specifically, the precursor frequency for allo-(H-2) antigens increased, while a decrease occurred in the frequency of anti-TNP-modified self clones. However, the total proportions of pCTL in the spleen was unchanged as judged by ConA stimulation of the cultures. Functionally, pCTL isolated from aged animals had a reduced proliferative capacity (decrease in burst size), normal specificity (no 3rd party lysis), but a reduced binding affinity for their targets. After transferring bone marrow cells of aged mice into irradiated young syngeneic recipients, it was demonstrated that the defects noted were an intrinsic property of the stem cells. Converse experiments were performed using young bone marrow stem cells transferred into an aged milieu; in this case the young cells also retained their characteristics. A further series of experiments examining the idiotypic repertoire of these CTL, demonstrated that the idiotypic receptors on cells from young mice were different from those of old mice and, furthermore, that the 'aged' repertoire was less heterogeneous and nonoverlapping with that of young mice. The results are in contrast with those obtained from the examination of syngeneic tumor-specific CTL [79], in which case the cells were found to lose their specificity with age and exhibited an altered idiotypic repertoire. It is therefore apparent that T cells in the aged animal, like that found for B cells, can express completely different idiotype/receptor repertoires from those found in young animals. Although the evidence presented above pertains to changes in the allotypic CTL repertoire with age, indirect evidence for similar changes in regulatory T cell subsets and their subsequent influence on the B cell repertoire has also been implied [74]. It appears that with age and with the appearance of an unspecified environmental influence(s) the T cell repertoire, at least for some specificities, undergoes selective pressure and eventually changes from that of the young.

Mucosal-Associated Lymphoid System

Without exception, studies have approached aging in terms of examining and describing changes solely within the systemic immune system (spleen, peripheral lymph nodes, and blood). While information in this regard is clearly important, the mucosal lymphoid apparatus, encompassing over a third of the bodies' lymphoid tissue, and forming the first line of defense in infection and foreign antigen exposure, has largely been ignored. This system has been shown to play a critical regulatory role in preferentially exposing the systemic immune system to selected antigens, while conferring resistance to others. It is comprised of a unique population of cells which exhibit distinctive gut homing properties, and characteristics. Evidence is also accumulating that the unique mucosal environment is capable of maintaining the growth and maturation of cells distinct from those in systemic areas, supporting the idea that this immune system while in communication with the rest of the animal can also remain compartmentalized and distinct.

The immune response in the mucosa is unique in the fact that it involves antibody mainly of the IgA isotype [80–82], under certain conditions can induce systemic unresponsiveness (oral tolerance) [83], stimulate reactivity solely within the mucosa [84], or evoke systemic responses as well [85–87]. The growing body of evidence dealing with gut immunity implicates Peyer's patches (PP) in all three of these phenomena [84, 88–94]. Such a finding is not unexpected since PP act as the entry point for much of the antigens contacting the gut, and furthermore, contains subpopulations of B cells [95] and Lyt-1^+ T lymphocytes which effect unique gut-related functions. The latter cell population is primarily involved in helping to generate specific IgA precursor plasma cells which mature further upon systemic migration and seeding of the gut mucosa [82, 88, 91].

Further information with regard to the unusual function of this tissue comes from a study performed by *Kiyono* et al. [88] who demonstrated that while PP have the full complement of cells required to generate an immune response in vitro (including functional antigen presenting cells [88, 96, 97; although not found in ref. 98], the tissue architecture in vivo prevents local immune response development. PP therefore appear to be essentially a factory producing activated precursor plasma cells which through their special characteristics and homing capabilities, return and generate immune responses within the gut tissue. To effect this unique function, PP contain at least two types of specialized T cells; one population of IgA FcR$^+$ cells which mediate a class switch in IgM and IgG bearing B cells to IgA [91, 92], and a second population which induces further maturation of these cells specifically [99, 100] or nonspecifically [82]. It has been speculated that the novel structure of PP germinal centers with their high content of Lyt-1^+ T cells and B cells expressing IgA, IgG, or IgM (but not IgD), allow for this class switch and the generation of memory and precursor IgA plasma cells, but limits maturation to this point [95].

PP are also exquisitely sensitive to the dose and/or route of antigen presentation for either triggering immune responsiveness or unresponsiveness [93, 101, 102]. Tolerance generated after oral ingestion of antigen can be mediated through suppressor T cells [85, 89, 90, 93, 102, 103], B cells [94, 101, 104, 105] or immune complexes [106]. Upon activation of cells within the PP, seeding of distant systemic sites with antigen-responsive cells may [84–87] or may not occur [84]. Conversely, intraperitoneal administration of antigen can cause a mucosal response to be detected [102] or not [101, 107]. Mixing of the immunization protocols (parenteral/oral) to generate a secondary response in the gut has been successful [107] or shown to induce suppression instead [101, 102]. It appears therefore that complex interactions occur between the systemic and mucosal lymphoid systems in generating an immune response.

Compartmentalization of the immune system within the mucosa is demonstrated by the following findings: *Gearhart and Cebra* [108] have shown that differences exist in the B cell repertoires directed against phosphorylcholine in the PP and spleen of nonimmune mice with the spleen containing larger numbers of cells bearing the TEPC-15 idiotype. Indirect support for this concept also comes from a study performed by *Jackson and Mestecky* [101] who demonstrated that after using a mixed immunization protocol (oral and intravenous administration of antigen) of bovine serum albumin, suppressive anti-idiotypic plasma cells could be detected in mucosal areas, but not after using either immunization route alone. It is speculated that the mixed protocol used in these experiments causes recruitment of cells bearing different idiotype repertoires, which circulate and prime for an anti-idiotypic response in the opposing area. It is evident from studies [107] showing cooperation when using this mixed immunization protocol that overlap in the repertoires must exist in some cases. Regardless, the strongest evidence for compartmentalization within the mucosa comes from the elegant study of *Eldridge* et al. [109] who demonstrated that an entirely different B cell population can exist in the mucosa of the X-linked immunodeficient (XID) CBA/N mouse, without any evidence of this population existing in systemic areas. The XID mutation results in a defective maturation of B cells, and an inability of CBA/N mice to respond to T-independent group 2 antigens. These investigators have found that the B cell subpopulation missing in the spleen of these animals, can be isolated from their PP at 6–8 weeks of age. The cells were capable of responding to T-independent group 2 and T-dependent antigens, and expressed the surface phenotype of mature B cells. Therefore, the available information supports not only distinct populations of isotype-specific T and B cells, but also cell subpopulations expressing unique receptor/idiotype repertoires within the mucosa [112, 113]. Whether this is a result of the gut environment with its incessant antigen exposure or regulatory interactions, remains unknown.

*Lack of Immune Dysfunction in Mucosal-Associated
Lymph Nodes with Age*

We have looked at the effect of aging on immune reactivity to TNP bovine gammaglobulin (TNP-BGG) immunization in C57BL/6J male mice. The magnitude of the immune response in old and young mice was measured in various lymphatic sites using the plaque-forming cell assay as a probe for changes in humoral immune reactivity at a cellular level. Age-kinetic studies revealed that the number of IgM, IgG and IgA anti-TNP PFC in the spleen reached a peak response at 4 months of age [110]. After 4 months, there was a significant decline in the number of PFC as compared with the peak response of mice at 4 months of age. We have also shown that old mice given antigen in the footpads and the base of the tail produced a reduced number of IgM, IgG and IgA PFC responses in the draining peripheral lymph nodes [110]. These results strongly supported the view that the systemic lymphoid system declines in immune function with age after parenteral administration of antigen.

In contrast, no significant decline in the number of IgM, IgG and IgA anti-TNP PFC in the mesenteric and mediastinal (or bronchial) lymph nodes of mice over 4 months of age was observed [110]. When TNP-BGG in complete Freund's adjuvant is given via gastric intubation, the anti-TNP PFC responses in the draining mesenteric lymph nodes as well as in the spleen from old mice were significantly enhanced as compared with the responses of the young group [110]. It is interesting that splenic PFC responses of old mice to TNP-BGG immunized via gastric intubation were in contrast to the findings when antigen is presented intraperitoneally. The exact nature of these differences due to route of immunization remains unclear at present. The findings in these studies suggest a lack of age-associated immune dysfunction in the mucosal-associated lymph nodes. This differential effect of aging on immune responses in mucosal and systemic lymphoid tissues in vivo suggests a site preference for an age-related decline in immune function.

*Heterogeneity of Affinity of the Primary PFC Response in
Mucosal Lymph Nodes of Different Aged Mice*

Male C57BL/6J mice of different ages were immunized intraperitoneally with TNP-BGG in CFA. The number of IgM, IgG and IgA anti-TNP PFC and the distribution of the PFC with regard to avidity in the spleen, mesenteric and mediastinal lymph nodes were determined 2 weeks after immunization. Avidity was determined by hapten inhibition of plaque formation. PFC from all lymphoid tissues of 2-month-old mice were highly heterogeneous and were of high average avidity. With age, the primary splenic PFC response was found to be markedly restricted in heterogeneity, with a preferential loss of IgM, IgG and IgA high avidity PFC [14]. The draining peripheral lymph node IgG PFC response of old mice exhibited a similar restriction in heterogeneity, with a preferential loss of high avidity PFC [14]. In contrast, mesenteric and mediastinal lymph node PFC responses of old mice remained highly heterogeneous with respect to antibody affinity in the three isotypes studied [14]. The results of these studies imply that T cell helper activity is unimpaired in the mucosal-associated lymphoid system of old mice, since it has been suggested [117] that impaired T cell helper activity may contribute to the preferential loss of high affinity PFC. The findings provide further support for the concept that the mucosal-associated lymphoid system differs from the systemic system with regard to its immune competence with age.

Auto-Anti-Idiotypic Antibody Regulation during Aging

Auto-anti-idiotypic antibodies have been shown to be responsible for a decline in the number of PFC, a reduction in the heterogeneity of avidity of PFC and the occurrence of anti-idiotype-blocked, hapten-augmentable PFC [111]. This auto-anti-idiotypic antibody, produced during the course of the immune response was explained, perhaps, by combining with B cell surface antigen receptors and thus inhibiting antibody secretion. Recently, we have demonstrated that the decline in the number of PFC, loss of high-affinity PFC and occurrence of hapten-augmentable PFC with age in the splenic B cell population was due to auto-anti-idiotypic antibody regulation [15]. This was subsequently confirmed by *Goidl* et al. [46]. Therefore, we examined the hypothesis that the lack of age-associated immune dysfunction in the mucosal-associated lymph nodes may be due to an absence of anti-idiotype-blocked, hapten-augmentable PFC immunoregulation. It was found that the mesenteric and mediastinal lymph node PFC responses of old mice exhibited no appreciable appearance of anti-idiotype-blocked, hapten-augmentable anti-TNP PFC in contrast to that found in the spleen [15]. Furthermore, mice receiving antigen in the footpads and the base of the tail also produced a significantly high percentage of hapten-augmentable PFC in the draining peripheral lymph nodes [112]. These results are consistent with the hypothesis that a lack of age-associated auto-anti-idiotypic antibody regulation occurs in the mucosal-associated lymph nodes of C57BL/6J mice. On the other hand, auto-anti-idiotypic antibody immunoregulation was found to be prevalent in the spleen and draining peripheral lymph nodes of the same old animals. These findings introduce an important issue, namely a division of the immune system into regulatory compartments.

Conclusions

Unfortunately, few studies have examined changes within the mucosal immune system during aging or have compared these changes with those in the systemic immune system [reviewed in 113]. In examining immune responses to TNP-BGG in splenic, peripheral lymph node (PLN), mediastinal lymph node (BLN), and mesenteric lymph node (MLN) areas, we have found that while systemic (spleen and PLN) IgM, IgG, and IgA PFC responses decline with age, mucosal PFC responses (MLN and BLN) remain vigorous in cells bearing all three isotypes [110]. This unimpaired mucosal response has also been observed by *Rivier* et al. [114], who found that IgA responses to the $\alpha(1-3)$-glucan determinant on dextran B1355 increased dramatically with age in the MLN. In that study, however, the splenic IgA immune response did not decline over the age span, but this may be a result of only testing 17-month-old Balb/C mice (middle age).

Furthermore, we have demonstated that: (a) the heterogeneity of a primary anti-TNP splenic and PLN PFC response became restricted during aging, with a preferential loss of high affinity clones; this restriction was not found in mucosal areas [14]; and (b) the appearance of anti-idiotype-blocked, hapten-augmentable PFC in the spleen and peripheral lymph nodes of aged mice, could not be demonstrated in the MLN and BLN of these animals [112]. This last study showed that antigen-specific B cells in systemic areas were selectively suppressed by anti-idiotypic antibody bound to their surface, but that none could be demonstrated on lymphocytes isolated from the mucosa of these same animals. The findings therefore strongly indicate that the idiotypic repertoires expressed by systemic and mucosal lymphocytes of aged mice are different and nonoverlapping. Furthermore, the inductive signals or idiotype/receptor changes responsible for the development of augmented systemic anti-idiotypic regulation during senescence [15, 18, 46, 64, 112, 115] are not triggered in the mucosal immune system [112] in response to either footpad or intraperitoneal injection of antigen.

The loss of high affinity plaque-forming cells and the reduced heterogeneity of the aged systemic immune response [14] could be accounted for by either the idiotype/receptor repertoire shift or impaired helper cell activity [117] previously described in systemic areas of aged animals [18, 24, 26, 35, 45, 46, 48, 56, 57, 63, 67, 68]. Moreover, the implied existence of a nonoverlapping B cell repertoire and functional T helper cell in the mucosa of these aged mice is supported by recent demonstrations of compartmentalization of the mucosal immune system [101, 108, 109].

The idea that the immune system of the gut remains vigorous and compartmentalized with age is exciting, since this area presents the first line of defence against many antigens encountered in the environment and would therefore protect aged individuals who's systemic immune responses have declined. In addition, the finding of two distinct immunological compartments in the body, one with an impaired and the other with an unimpaired immune circuitry, would allow further insight into immune regulation, its change with age, and eventually might permit a possible means of boosting systemic immune responses in aged individuals with disseminated infections or cancer.

References

1 Friedberger, E.; Bock, G.; Furstenheim, A.: Zur Normalantikörperkurve des Menschen durch die verschiedenen Lebensalter und ihre Bedeutung für die Erklärung der Hautteste (Schick Dick). Z. ImmunForsch. exp. Ther. *64:* 294–319 (1929).

2 Thomsen, O.; Kettel, K.: Die Stärke der menschlichen Isoagglutinine und entsprechenden Blutkörperchenrezeptoren in verschiedenen Lebensaltern. Z. ImmunForsch. exp. Ther. *63:* 67–93 (1929).

3 Boyd, E.: The weight of the thymus gland in health and in disease. Am. J. Dis. Child. *43:* 1162–1214 (1932).

4 Santisteban, G.A.: The growth and involution of lymphatic tissue and its interrelationship to aging and to the growth of the adrenal glands and sex organs in CBA mice. Anat. Res. *136:* 117–126 (1960).

5 Smith, G.S.; Walford, R.L.: Influence of the main histocompatibility complex of ageing in mice. Nature, Lond. *270:* 727–729 (1977).

6 Wallace, D.J.; Bluestone, R.; Klironberg, J.R.: The biology of aging and the immune system. Bull. Rheum. Dis. *32:* 13–19 (1982).

7 Good, R.A.; Yunis, E.J.: Association for autoimmunity, immunodeficiency and aging in man, rabbits and mice. Fed. Proc. *33:* 2040–2050 (1974).

8 Stobo, J.D.; Tomasi, T.B.: Aging and the regulation of immune reactivity. J. clin. Dis. *28:* 437–440 (1975).
9 Makinodan, T.; Heidrich, M.L.; Nordin, A.A.: Immunodeficiency and autoimmunity in aging. Birth Defects *11:* 193–198 (1975).
10 Weksler, M.E.: The senescence of the immune system. Hosp. Pract. *16:* 53–64 (1981).
11 Dupere, S.L.F.; Kolodziej, B.J.: Cellular and molecular aspects of thymus dependent antibody production in aged C3H/HeBr mice. Age *6:* 11–19 (1983).
12 Doggett, D.L.; Chang, M.P.; Makinodan, T.; Strehler, B.L.: Cellular and molecular aspects of immune system aging. Mol. clin. Biochem. *37:* 137–156 (1981).
13 Weksler, M.E.: The immune system and the aging process in man. Proc. Soc. exp. Biol. Med. *165:* 200–205 (1980).
14 Szewczuk, M.R.; Campbell, R.J.: Differential effect of aging on the heterogeneity of the immune response to a T dependent antigen in systemic and mucosal-associated lymphoid tissues. J. Immun. *126:* 472–477 (1981).
15 Szewczuk, M.R.; Campbell, R.J.: Loss of immune competence with age may be due to auto-anti-idiotypic antibody regulation. Nature, Lond. *286:* 164–166 (1980).
16 Frasca, D.; Garavini, M.; Doria, G.: Recovery of T-cell functions in aged mice infected with synthetic thymosin 1. Cell. Immunol. *72:* 384–391 (1982).
17 Harrison, D.E.: Long-term erythropoietic repopulating ability of old, young, and fetal stem cells. J. exp. Med. *157:* 1496–1504 (1983).
18 Goidl, E.A.; Choy, J.W.; Gibbans, J.J.; Weksler, M.E.; Thorbecke, G.J.; Siskind, G.W.: Production of auto-anti-idiotypic antibody during the normal immune response. VII. Analysis of the cellular basis for the increased auto-anti-idiotypic antibody production by aged mice. J. exp. Med. *157:* 1635–1645 (1983).
19 Weksler, M.E.; Innes, J.B.; Goldstein, G.: Immunological studies of aging. IV. The contribution of thymic involution to the immune deficiencies of aging mice and reversal with thymopoietin$_{32-36}$. J. exp. Med. *148:* 996–1006 (1978).
20 Makinodan, T.; Kay, M.M.B.: Age influence on the immune system. Adv. Immunol. *29:* 287–330 (1980).
21 Price, G.B.; Makinodan, T.: Immunological deficiencies in senescence. I. Characterization of intrinsic deficiencies. J. Immun. *108:* 403–412 (1972).
22 Price, G.B.; Makinodan T.: Immunological deficiencies in senescence. II. Characterization of extrinsic deficiencies. J. Immun. *108:* 413–417 (1972).
23 Kay, M.M.B.: The thymus. Clock for immunological aging? J. invest. Derm. *73:* 29–38 (1979).
24 Goidl, E.A.; Innes, J.B.; Weksler, M.E.: Immunological studies of aging. II. Loss of IgG and high avidity plaque-forming cells and increased suppressor cell activity in aging mice. J. exp. Med. *144:* 1037–1048 (1976).
25 Krogsrud, R.L.; Perkins, E.H.: Age related changes in T cell function. J. Immun. *118:* 1607–1611 (1977).
26 Callard, R.E.; Basten, A.: Immune function in aged mice. IV. Loss of T cell and B cell function in thymic dependent antibody responses. Eur. J. Immunol. *8:* 552–558 (1978).
27 Callard, R.E.; Basten, A.; Waters, L.K.: Immune function in aged mice. II. B cell function. Cell. Immunol. *31:* 26–36 (1977).
28 Amagai, T.; Nakano, K.; Cinader, B.: Mechanism involved in age-dependent decline of immune responsiveness and apparent resistance against tolerance induction in C57BL/6 mice. Scand. J. Immunol. *16:* 217–231 (1982).

29 Wilson, D.A.; Braley-Mullen, H.: Immunoregulation in MRL/Mp-Ipr/Ipr mice: Evidence for decreased helper T cell and increased suppressor T cell function with age. Cell. Immunol. *74:* 72–85 (1982).
30 Makinodan, T.; Albright, J.W.; Good, P.I.; Peter, C.P.; Heidrick, M.C.: Reduced humoral immune activity in long lived old mice. An approach to elucidate its mechanisms. Immunology *31:* 903–911 (1976).
31a Callard, R.E.; Fazekas de St Groth, B.; Basten, A.; McKenzie, I.F.C.: Immune function in aged mice. V. Role of suppressor cells. J. Immun. *124:* 52–58 (1980).
31b Roder, J.R.; Duwe, A.K.; Bell, D.A.; Singhal, S.K.: Immunological senescence. I. The role of suppressor cells. Immunology *35:* 837–848 (1978).
32 Jaroslow, B.N.; Suhrbier, K.M.; Fry, R.J.M.; Tyler, S.A.: In vitro suppression of immunocompetent cells by lymphomas from aging mice. J. natn. Cancer Inst. *541:* 1427–1432 (1975).
33 Dekruyff, R.H.; Kim, Y.T.; Siskind, G.W.; Weksler, M.E.: Age-related changes in the in vitro immune response: increased suppressor activity in immature and aged mice. J. Immun. *125:* 142–147 (1980).
34 Rosenberg, J.S.; Gilman, S.C.; Feldman, J.D.: Effects of aging on cell cooperation and lymphocyte responsiveness to cytokines. J. Immun. *130:* 1754–1758 (1983).
35 Doria, G.; D'Agostaro, G.; Garavini, M.: Age-dependent changes of B-cell reactivity and T cell-T cell interaction in the in vitro antibody response. Cell. Immunol. *53:* 195–206 (1980).
36 Ceuppens, J.L.; Goodwin, J.S.: Regulation of immunoglobulin production in pokeweed mitogen stimulated cultures of lymphocytes from young and old adults. J. Immun. *128:* 2439–2434 (1982).
37 Weiner, H.L.; Moorhead, J.W.; Claman, H.W.: Anti-immunoglobulin stimulation of murine lymphocytes. I. Age-dependence of the proliferative response. J. Immun. *116:* 1656–1661 (1976).
38 Scribner, D.J.; Moorhead, J.W.: Anti-immunoglobulin stimulation of murine lymphocytes. VII. Identification of an age-dependent change in accessory cell function. J. Immun. *128:* 1377–1380 (1982).
39 Taylor, R.B.: Regulation of antibody responses by antibody towards the immunogen. Immunol. Today *3:* 47–51 (1982).
40 Morgan, E.L.; Thoman, M.L.; Weigle, W.D.: The immune response in aged C57BL/6J mice. I. Assessment of lesions in the B-cell and T-cell compartments of aged mice utilizing the Fc fragment mediated polyclonal antibody response. Cell. Immunol. *63:* 16–27 (1981).
41 Woda, B.A.; Yguerabide, J.; Feldman, J.D.: Mobility and density of Ag B, Ia and Fc receptors on the surface of lymphocytes from young and old rats. J. Immun. *123:* 2161–2167 (1979).
42 Rosenberg, J.S.; Gilman, S.C.; Feldman, D.J.: Activation of rat B lymphocytes. II. Functional and structural changes in 'aged' rat B lymphocytes. J. Immun. *128:* 656–660 (1982).
43 Woda, B.A.; Feldman, J.D.: Density of surface immunoglobulin and capping on rat B lymphocytes. I. Change with age. J. exp. Med. *149:* 416–423 (1979).
44 Wagner, A.P.; Wagner, L.P.; Psarrou, E.: Age-related changes in 28s and 18s RNA in antigen-stimulated and non-stimulated rat spleens. Age *5:* 113–117 (1982).
45 Morgan, E.L.; Weigle, W.O.: The immune response in aged C57BL/6J mice. II. Characterization and reversal of a defect in the ability of aged spleen cells to respond to the adjuvant properties of Fc fragments. J. Immun. *129:* 36–39 (1982).

46 Goidl, E.A.; Thorbecke, G.J.; Weksler, M.E.; Siskind, G.W.: Production of auto anti-idiotypic antibody during the normal immune response: changes in the auto anti-idiotypic antibody response and the idiotypic repertoire associated with aging. Proc. natn. Acad. Sci. USA 77: 6788–6792 (1980).
47 Abe, T.; Morimoto, C.; Toguchi, T.; Kiyotaki, M.; Homma, M.: Evidence of aberation of T-cell subsets in aged individuals. Scand. J. Immunol. 13: 151–157 (1981).
48 Liu, J.J.; Segre, M.; Segre, D.: Changes in suppressor, helper and B-cell functions in aging mice. Cell. Immunol. 66: 372–382 (1982).
49 Smith, A.M.: The effects of age on the immune response to type III pneumococcal polysaccharide (SIII) and bacterial lipopolysaccharide (LPS) in BALB/c, SJL/J, and C3H mice. J. Immun. 116: 469–474 (1976).
50 Makinodan, T.; Adler, W.H.: Effects of aging on the differentiation and proliferation potential of cells of the immune system. Fed. Proc. 34: 153–158 (1975).
51 Chen, M.G.: Age-related changes in hematopoietic stem cell populations of a long-lived hybrid mouse. J. cell. Physiol. 78: 225–232 (1971).
52 Harrison, D.E.; Doubleday, J.W.: Normal function of immunologic stem cells from aged mice. J. Immun. 114: 1314–1317 (1975).
53 Callard, R.E.: Immune function in aged mice. III. Role of macrophages and effect of 2-mercaptoethanol in the response of spleen cells from old mice to phytohemagglutinin, lipopolysaccharide and allogeneic cells. Eur. J. Immunol. 8: 697–705 (1978).
54 Perkins, E.H.; Makinodan, T.: Nature of humoral immunologic deficiencies of the aged. Proc. Rocky Mountain Symp. Aging, Colorado State Univ., Fort Collins, 1971, p. 80.
55 Heidrick, M.L.: Age-related changes in hydrolase activity of peritoneal macrophages. Gerontologist 12: 28 (1972).
56 Chang, M.P.; Makinodan, T.; Peterson, W.J.; Strehler, B.L.: Role of T cells and adherent cells in age related decline in interleukin 2 production. J. Immun. 129: 2436–2430 (1982).
57 Thoman, M.L.; Weigle, W.O.: Cell-mediated immunity in aged mice with an underlying lesion in IL2 synthesis. J. Immun. 128: 2358–2361 (1982).
58 Gorczynski, R.M.; Kennedy, M.; MacRay, S.: Alteration in lymphocyte recognition repertoire during aging. II. Changes in the expressed T-cell receptor repertoire in aged mice and the persistence of that change after transplantation to a new differentiative environment. Cell. Immunol. 75: 226–241 (1983).
59 Roberts-Thomson, I.C.; Whittingham, S.; Youngchaiyud, U.; MacKay, I.R.: Aging, immune response and mortality. Lancet ii: 368–370 (1974).
60 Patel, P.J.: Aging and antimicrobial immunity. Impaired production of mediator T cells as a basis for the decreased resistance of senescent mice to listeriosis. J. exp. Med. 154: 821–831 (1981).
61 Bentley, D.M.; Morris, R.E.: T cell subsets required for protection against age-dependent polioencephalomyelitis of C58 mice. J. Immun. 128: 530–534 (1982).
62 Zuberi, R.I.; Katz, D.H.: Genetics of cell interactions in aged mice: age-related decline in capacity of parental cells or environment to induce allogeneic effects on F1 lymphocytes. J. Immun. 129: 272–277 (1982).
63 Thoman, M.L.; Weigle, W.O.: Lymphokines and aging. Interleukin-2 production and activity in aged animals. J. Immun. 127: 2102–2106 (1981).
64 Klinman, N.R.: Antibody-specific immunoregulation and the immunodeficiency of aging. J. exp. Med. 154: 547–551 (1981).

65 Segre, D.; Segre, M.: Humoral immunity in aged mice. II. Increased suppressor T cell activity in immunologically deficient old mice. J. Immun. *116:* 735–738 (1976).
66 Gillis, S.; Kozak, R.; Durante, M.; Weksler, M.E.: Immunological studies of aging. Decreased production of and response to T cell growth factor by lymphocytes from aged humans. J. clin. Invest. *67:* 937–942 (1981).
67 Szewczuk, M.R.: Synergistic cooperation between T and B lymphocytes from old mice in the production of auto-anti-idiotypic antibody regulation in adult irradiated hosts. Can. J. Aging *1/2:* 3–10 (1982).
68 Gorczynski, R.M.; Kennedy, M.; MacRae, S.; Benzing, K.; Price, G.B.: Alterations in lymphocyte recognition repertoire during ageing. I. Analysis of changes in immune response potential of B lymphocytes from non-immunized aged mice, and the role of accessory cells in the expression of that potential (submitted for publication, 1983).
69 Kishimoto, S.; Takahama, T.; Mizumachi, H.: In vitro immune response to the 2,4,6-trinitrophenylated determinant in aged C57BL/6J mice: changes in the humoral immune respone to, avidity for the TNP-determinant and the responsiveness to LPS, effect with age. J. Immun. *116:* 294–300 (1976).
70 Cramer, M.; Krawinkel, U.; Metcher, I.; Imanishi-Kari, J.; Ben-Neriah, V.; Givol, D.; Rajewski, K.: Isolated hapten-binding receptors of sensitized lymphocytes. IV. Expression of immunoglobulin variable regions in (4-hydroxy-3-nitrophenyl) acetyl (NP)-specific receptors isolated from murine B and T lymphocytes. Eur. J. Immunol. *9:* 332–338 (1979).
71 Woodland, R.; Cantor, H.: Idiotype-specific T helper cells are required to induce idiotype-positive B memory cells to secrete antibody. Eur. J. Immunol. *8:* 600–606 (1978).
72 Gorczynski, R.M.; Khomasurya, B.; Kennedy, M.; MacRae, S.; Cunningham, A.J.: Individual-specific (idiotypic) T-B interactions regulating the production of anti-2,4,6-trinitrophenyl antibody. II. Development of idiotype-specific helper and suppressor T cells within mice making an immune response. Eur. J. Immunol. *10:* 788–791 (1980).
73 Gorczynski, R.M.; Cunningham, A.J.: Requirement for matching T cell and B cell subsets in secondary anti-hapten antibody responses. Eur. J. Immunol. *8:* 753–755 (1978).
74 Zharhary, D.; Klinman, N.R.: Antigen responsiveness of the mature and generative B cell populations of aged mice. J. exp. Med. *157:* 1300–1308 (1983).
75 Kay, M.M.B.: Parainfluenza infection of aged mice results in autoimmune disease. Clin. Immunol. Immunopath. *12:* 301–315 (1979).
76 Larsson, E.; Coutinho, A.; Martinez-A., C.: A suggested mechanism for T lymphocyte activation: implications on the acquisition of functional reactivities. Immunol. Rev. *51:* 61–91 (1980).
77 Watson, J.: Continuous proliferation of murine antigen specific helper T lymphocytes in culture. J. exp. Med. *150:* 1510–1519 (1979).
78 Howard, M.; Mizel, S.B.; Lachman, L.; Ansel, J.; Johnston, B.; Paul, W.E.: Role of interleukin 1 in anti-immunoglobulin induced B cell proliferation. J. exp. Med. *157:* 1529–1543 (1983).
79 Flood, P.M.; Urban, J.L.; Kripke, M.L.; Schreiber, H.: Loss of tumor-specific and idiotype specific immunity with age. J. exp. Med. *154:* 275–290 (1981).
80 Tseng, J.: Expression of immunoglobulin isotypes by lymphoid cells of mouse intestinal lamina propria. Cell. Immunol. *73:* 324–336 (1982).
80 Komisar, J.L.; Fuhrman, J.A.; Cebra, J.J.: IgA-producing hybridomas are readily derived from gut-associated lymphoid tissue. J. Immun. *128:* 2376–2378 (1982).

82 Tseng, J.: Expression of immunoglobulin heavy chain isotypes by Peyer's patch lymphocytes stimulated with mitogens in culture. J. Immun. *128:* 2719–2725 (1982).
83 Challocombe, S.J.; Tomasi, T.B., Jr.: Systemic tolerance and secretory immunity after oral immunization. J. exp. Med. *152:* 1459–1472 (1980).
84 Muller-Schoop, J.W.; Good, R.A.: Functional studies of Peyer's patches. Evidence for their participation in intestinal responses. J. Immun. *114:* 1757–1760 (1975).
85 Dolezel, J.; Bienenstock, J.: γ-A and non-γ-A immune response after oral and parenteral immunization of the hamster. Cell. Immunol. *2:* 458–468 (1971).
86 Dolezel, J.; Bienenstock, J.: Response of the hamster to oral and parenteral immunization. Cell. Immunol. *2:* 326–334 (1971).
87 Rothberg, R.M.; Kraft, S.C.; Farr, R.S.: Similarities between rabbit antibodies produced following ingestion of bovine serum albumin and following parenteral immunization. J. Immun. *98:* 386–395 (1967).
88 Kiyono, H.; McGhee, J.R.; Wannemuehler, M.J.; Frangakis, M.V.; Spalding, D.M.; Machalek, S.M.; Koopman, W.J.: In vitro immune response to a T cell-dependent antigen by cultures of disassociated murine Peyer's patches. Proc. natn. Acad. Sci. USA *79:* 596–600 (1982).
89 Mattingly, J.A.; Waksman, B.H.: Immunologic suppression after oral administration of antigen. I. Specific suppressor cells formed in rat Peyer's patches after oral administration of sheep erythrocytes and their systemic migration. J. Immun. *121:* 1878–1883 (1978).
90 Ngan, J.; Kind, L.S.: Suppressor T cells for IgE and IgG in Peyer's patches of mice made tolerant by the oral administration of ovalbumin. J. Immun. *120:* 861–865 (1978).
91 Kawanishi, H.; Saltzman, L.E.; Strober, W.: Characteristics and regulatory function of murine Con A-induced, cloned T cells obtained from Peyer's patches and spleen: mechanisms regulating isotype-specific immunoglobulin production by Peyer's patch B cells. J. Immun. *129:* 475–483 (1982).
92 Kawanishi, H.; Saltzman, L.E.; Strober, W.: Mechanisms regulating IgA class-specific immunoglobulin production in murine gut associated lymphoid tissues. J. exp. Med. *157:* 433–450 (1983).
93 MacDonald, T.T.: Immunosuppression caused by antigen feeding. II. Suppressor cells mask Peyer's patch B cell priming to orally administered antigen. Eur. J. Immunol. *13:* 138–142 (1983).
94 Asherson, G.L.; Zembala, M.; Perera, M.A.C.C.; Mayhew, B.; Thomas, W.R.: Production of immunity and unresponsiveness in the mouse by feeding contact sensitizing agents and the role of suppressor cells in the Peyer's patches, mesenteric lymph node, and other lymphoid tissue. Cell. Immunol. *33:* 145–155 (1977).
95 Butcher, E.C.; Rouse, R.V.; Coffman, R.L.; Nottenburg, C.N.; Hardy, R.R.; Weisman, I.C.: Surface phenotype of Peyer's patch germinal center cells. Implications for the role of germinal centers in B cell differentiation. J. Immun. *129:* 2698–2706 (1982).
96 Sminia, T.; Wilders, M.M.; Janse, E.M.; Hoelsmit, E.C.M.: Characterization of non-lymphoid cells in Peyer's patches of the rat. Immunobiology *164:* 136–143 (1983).
97 Spalding, D.M.; Koopman, W.J.; Eldridge, J.H.; McGhee, J.R.; Steinman, R.M.: Accessory cells in murine Peyer's patches. I. Identification and enrichment of a functional dendritic cell. J. exp. Med. *157:* 1646–1659 (1983).
98 Kagnoff, M.F.: Functional characteristics of Peyer's patch cells. III. Carrier priming of T cells by antigen feeding. J. exp. Med. *142:* 1425–1435 (1975).

99 Elson, C.O.; Heck, J.A.; Strober, W.: T-cell regulation of murine IgA synthesis. J. exp. Med. *149:* 632–643 (1979).
100 Kiyono, H.; McGhee, J.R.; Mosteller, L.M.; Eldridge, J.H.; Koopman, W.J.; Kearney, J.F.; Michalek, S.M.: Murine Peyer's patch T cell clones. Characterization of antigen-specific helper T cells for immunoglobulin-A response. J. exp. Med. *156:* 1115–1130 (1982).
101 Jackson, S.; Mestecky, J.: Oral-parenteral immunization leads to the appearance of IgG auto-anti-idiotypic cells in mucosal tissues. Cell. Immunol. *60:* 498–502 (1981).
102 Pierce, N.F.; Koster, F.T.: Priming and suppression of the intestinal immune response to cholera toxoid/toxin parenteral toxoid in rats. J. Immun. *124:* 307–311 (1980).
103 Richman, L.K.; Chiller, J.M.; Brown, W.R.; Harrison, D.G.; Vaz, N.M.: Enterically induced immunological tolerance. I. Induction of suppressor T lymphocytes by intragastric administration of soluble proteins. J. Immun. *121:* 2429–2434 (1978).
104 Kagnoff, M.F.: Effect of antigen feeding on intestinal and systemic immune responses. III. Antigen-specific serum-mediated suppression of humoral antibody responses after antigen feeding. Cell. Immunol. *40:* 186–203 (1978).
105 Hanson, D.G.; Miller, S.D.: Inhibition of specific immune responses by feeding protein antigens. V. Induction of the tolerant state in the abcence of specific suppressor T cells. J. Immun. *128:* 2378–2381 (1982).
106 André, C.; Heremans, J.F.; Vaerman, J.P.; Cambiaso, C.L.: A mechanism for the induction of immunological tolerance by antigen feeding. Antigen-antibody complexes. J. exp. Med. *142:* 1509–1519 (1975).
107 Pierce, N.F.; Gowans, J.L.: Cellular kinetics and the intestinal immune response to cholera toxoid in rats. J. exp. Med. *142:* 1550–1563 (1975).
108 Gearhart, P.J.; Cebra, J.J.: Differentiated B lymphocytes. Potential to express particular antibody variable and constant regions depends on site of lymphoid tissue and antigen load. J. exp. Med. *149:* 216–227 (1979).
109 Eldridge, J.H.; Kiyono, H.; Michalek, S.M.; McGhee, J.R.: Evidence for a mature B cell subpopulation in Peyer's patches of young adult XID mice. J. exp. Med. *157:* 789–794 (1983).
110 Szewczuk, M.R.; Campbell, R.J.; Jung, L.K.: Lack of age-associated immune dysfunction in mucosal-associated lymph nodes. J. Immun. *126:* 2200–2204 (1981).
111 Schrater, A.F.; Goidl, E.A.; Thorbecke, G.J.; Siskind, G.W.: Production of auto-anti-idiotypic antibody during the normal immune response to TNP-Ficoll. I. Occurrence in AKR/J and BALB/c mice of hapten-augmentable, anti-TNP plaque-forming cells and their acceleration appearance in recipients of immune spleen cells. J. exp. Med. *150:* 138–153 (1979).
112 Szewczuk, M.R.; Campbell, R.J.: Lack of age associated auto auto-anti-idiotypic antibody regulation in mucosal-associated lymph nodes. Eur. J. Immunol. *11:* 650–656 (1981).
113 Szewczuk, M.R.; Wade, A.W.: Aging and the mucosal-associated lymphoid system. Ann. N.Y. Acad. Sci. *409:* 333–344 (1983).
114 Rivier, D.A.; Trefts, P.E.; Kagnoff, M.F.: Age-dependence of the IgA anti-α(1-3) dextran B1355 response in vitro. Scand. J. Immunol. *17:* 115–121 (1983).
115 Goidl, E.A.; Schrater, A.F.; Thorbecke, G.J.; Siskind, G.W.: Production of auto-anti-idiotypic antibody during the normal immune response. Eur. J. Immunol. *10:* 810–814 (1980).
116 Nishikawa, S-I.; Takemori, T.; Rajewsky, K.: The expression of a set of antibody variable

regions in lipopolysaccharide-reactive B cells at various stages of ontogeny and its control by anti-idiotype antibody. Eur. J. Immunol. *13:* 318–325 (1983).
117 Dekruyff, R.H.; Siskind, G.W.: Studies on the control of antibody synthesis. XIV. Role of T cells in regulating antibody affinity. Cell. Immunol. *47:* 134–142 (1979).

M.R. Szewczuk, PhD, Associate Professor, Department of Microbiology and Immunology, Queen's University, Kingston, Ont. K7L 3N6 (Canada)

Lymphocyte Differentiation and Its Microenvironment in the Human Thymus during Aging[1]

Gerhard G. Steinmann, Hans-Konrad Müller-Hermelink[2]

Pathologisches Institut der Universität Kiel, BRD

Introduction

Although the biological mechanisms of aging are poorly understood, many observations have suggested close relations between senescence and immunity. In particular, immune functions which depend on the T cell system decline with aging [for reviews see 14, 27]. Since involution of the thymus precedes the decline of T cell function and might act on a T cell differentiation pathway, the thymus is widely regarded as the most limiting factor of the immunologic capacity of the aged [10, 13]. Recent studies in mice as well as in man have elucidated the role of the thymus in the generation of functional subpopulations of T lymphocytes [4, 12, 21, 28]. Prothymocytes originate from bone marrow precursors and learn within the thymic microenvironment to recognize antigens of the major histocompatibility complex (MHC). Recognition of MHC antigens is obligatory for the recognition of non-self antigens by mature T lymphocytes and their selective sequestration in peripheral lymphoid organs [15, 23]. Stationary nonlymphoid cells include thymic nurse cells [26], cortical and medullary epithelial cells, as well as interdigitating reticulum cells (IDC) as major components [8, 25]. As shown by immunoelectron microscopy, cortical and medullary epithelial cells are the most important MHC-positive elements within the mouse thymus [7].

Our present immunohistologic study of lymphocyte differentiation and its thymic microenvironment during aging demonstrates a distinct heterogeneity of the framework of epithelial cells of the cortical thymus in man, which is already evident in biopsies of young healthy individuals. Different

---

[1] This work was supported by a DFG-grant (Ste 296/5-2).
[2] We thank Drs. *R. Evans* and *F.J. Bollum* for providing antibodies and Ms. *B. Zanger* for technical assistance.

phenotypes of cortical epithelium suggest sufficient and nonsufficient functions in relation to T cell maturation, and might lead to a hypothesis, how physiological involution of the thymus could begin. In contrast, immature T cell precursors seem to appear continuously in the thymus cortex during aging, and differentiation pathways seem to be preserved throughout the end of life.

Materials and Methods

We obtained biopsies of normal thymuses from patients undergoing cardiac surgery. A total of 102 samples from 41 persons were examined. Their ages ranged from 6 months to 71 years. Their immune system was in a healthy state as proven by a profound preoperative clinical examination.

Immunoperoxidase staining of frozen sections was performed as follows. Lyophilized 8-μm frozen sections were fixed in acetone for 10 min and subsequently in chloroform for 10 min [24], both at room temperature. Fixed sections were incubated at room temperature with dilutions of purified monoclonal antibodies (table I) ranging from 0.5 to 2 μg/ml for 1 h. Sections were then washed in phosphate-buffered saline (PBS) and treated with peroxidase-conjugated rabbit antimouse IgG (Dako, Copenhagen, Denmark) for 30 min to which normal human serum at a final dilution of 1:3 had been added in order to reduce nonspecific bindings. For control of nonspecific bindings, primary antibody was substituted by control ascites fluid (table I).

For demonstration of terminal transferase (TdT) a modified method of *Janossy* et al. [11] was applied. Briefly, sections from shock frozen tissue blocks were cut in a cryostat and immediately fixed in 10% buffered formalin for 15 min. After washing in PBS, endogenous peroxidase was blocked with 0.3% H_2O_2/methanol. Antibody incubations followed: (1) normal goat serum (Polyscience, Warrington, Pa.); (2) rabbit antibovine TdT (table I) which is a specific antibody to highly purified TdT, producing precipitates and neutralizing homologous antigen, and which cross-reacts with human terminal transferase [3]; (3) goat antirabbit IgG (Litton Bionetics, Kensington, Md.), and (4) peroxidase-anti-peroxidase from rabbit (Dako, Copenhagen, Denmark). Controls were performed substituting incubation 2 with normal rabbit serum (Polyscience). The specificity of the rabbit antibody for human TdT has been proven before by correlation with results of biochemical determinations [16, 17].

Subsequently, all sections were washed in PBS and stained with freshly prepared diaminobenzidine tetrahydrochloride solution (1 mg/ml, Walther, Kiel, FRG) supplemented with 0.03% H_2O_2. After washing in running tap water, slides were mounted either without counterstain or counterstained with hemalum.

Results

HLA-A,B,C and HLA-Dr

As shown in figure 1, subcapsular regions stained intensely for HLA-A,B,C and featured usually a tight network of peripheral staining surround-

Table I. Monoclonal and polyclonal antibodies used to identify immunologic characteristics of the human thymus in situ

Antibody	Clones or producer animal	Antigen or cells identified	Source
1 Anti-HLA	61D2	major serologically defined antigens HLA-A,B,C	Bethesda Research Laboratories, Neu-Isenburg, FRG
2 Anti-HLA-Dr	7.2	HLA-Dr antigens (Ia-like; p 28,33)	New England Nuclear, Dreireich, FRG
3 Anti-Leu-6	SK9	common thymocytes, IDCs, and Langerhans cells of the skin	Dr. *R. Evans,* New York, N.Y.
4 Anti-Leu-2a	SK1	suppressor/cytotocic subset of T cells	Dr. *R. Evans,* New York, N.Y.
5 Anti-Leu-3a	SK3	helper/inducer subset of T cells	Dr. *R. Evans,* New York N.Y.
6 Anti-Leu-4	SK7	sheep erythrocyte receptor-positive T cells (95%)	Becton-Dickinson, Sunnyvale, Calif.
7 Anti-Leu-1	L17F12	all T cells (gp 67)	Becton-Dickinson, Sunnyvale, Calif.
8 Anti-TdT	rabbit	terminal transferase, cortical thymocytes	Dr. *F. J. Bollum,* Bethesda, Md. Bethesda Research Laboratories, Neu-Isenburg, FRG
9 Control ascites	P3x63 myeloma	unknown	Bethesda Research Laboratories, Neu-Isenburg, FRG

ing lymphocytes and stationary cells independent of the age of the tissue donor. In contrast, the outermost epithelial layer was predominantly negative for HLA-Dr (fig. 2). The staining of the more central regions of the cortex exhibited a dominating pattern for HLA-A,B,C as well as for HLA-Dr. This pattern (type A) was built up by a tender reticular framework of positive reactions of the cortical epithelium stained for HLA-Dr, which resembled channels and tube-like formations with an axis approximately perpendicular to the outer capsule of the lobule. The loose network contained numerous HLA-Dr-negative cortical lymphocytes, which had contact to a HLA-Dr-

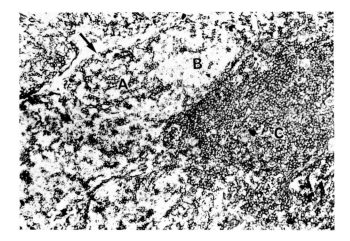

Fig. 1. Thymus, 5 years, male. Immunoperoxidase, anti-HLA-A,B,C. Arrow: positive outermost subcapsular region. Three types of staining patterns are shown: reticular (A), depleted (B) and confluent (C) – see text. Counterstained. × 70.

Fig. 2. Thymus, 9 months, female. Immunoperoxidase, anti-HLA-Dr. Arrow = Predominately negative outermost subcapsular layer. Note reticular pattern of epithelial cells within the cortex (c) and confluent staining pattern within the medulla (m). Counterstained. × 70.

Table II. Distribution of antigens as determined in frozen sections of thymus biopsies and alterations during aging (in parentheses)

Antigen	Cortex	Medulla	Perivascular space
1 HLA-A,B,C	heterogenous peripheral staining: (A) reticular pattern 80–97 %[1] (B) depleted pattern with faint peripheral staining of lymphocytes 3–20 % (decrease)	homogenous intense peripheral staining of most of the cells: (C) confluent pattern (no change)	homogenous slight peripheral staining of lymphoid cells (B) depleted pattern (increase)
2 HLA-Dr	heterogenous peripheral staining: (A) reticular pattern 70–95 % (increase) (B) depleted pattern with faint peripheral staining of lymphocytes 5–30 % (decrease)	homogenous peripheral staining of most of the cells: (C) confluent pattern (no change)	homogenous faint peripheral staining of some lymphoid cells (B) depleted pattern (increase)
3 Leu-6	intense peripheral staining, > 95 % (no change)	peripheral staining, 7–43 % (decrease)	peripheral staining, 7–19 % (no change)
4 Leu-2a	intense peripheral staining, 70–99 % (no change)	peripheral staining, 11–79 % (increase)	peripheral staining, 7–31 % (increase)
5 Leu-3a	intense peripheral staining, > 98 % (no change)	peripheral staining, 70–94 % (no change)	peripheral staining, 7–34 % (no change)
6 Leu-4	intense peripheral staining, > 98 % (no change)	peripheral staining, 89–94 % (no change)	peripheral staining, 9–34 % (decrease)
7 Leu-1	peripheral staining, 10–39 % (increase)	intense peripheral staining, 84–98 % (no change)	peripheral staining, 10–41 % (no change)
8 TdT	nuclear staining, > 95 % (no change)	no staining (no change)	no staining (no change)

[1] Percentages estimated using point counting and cell counting procedures

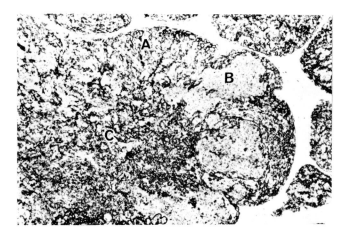

Fig. 3. Thymus, 8 years, male. Immunoperoxidase, anti-HLA-Dr. Three types of staining patterns: A = type A pattern: tender reticular framework of epithelial cells straightened perpendicularly in relation to the capsule of the lobule. B = Type B pattern: HLA-Dr-depleted areas with faintly stained lymphocytes. C = Type C pattern: confluent HLA-Dr reaction of medullary epithelial cells and lymphocytes. Counterstained. ×28.

positive structure. In comparison to HLA-Dr staining, HLA-A,B,C staining resulted in a similar network which seemed to be significantly more strongly interwoven. As age of the tissue donor increased, HLA-A,B,C positive cortical reticulums became increasingly interrupted. However, type A patterns of cortical staining for HLA-A,B,C or HLA-Dr could be easily identified in all biopsies of older persons examined (fig. 3).

A completely different staining pattern (type B) for HLA-Dr and HLA-A,B,C was observed in the neighborhood of the type A pattern. Type B patterns were characterized by a complete loss of reticular HLA-Dr-positive epithelial structures in smaller and larger islands within the cortex of the thymus. Many lymphocytes were aggregated without demonstrable contact to epithelial cells. The lymphocytes, here, exhibited a faint positive reaction for HLA-Dr instead (fig. 3). During aging, a small quantitative decrease of cortical tissue with type B pattern could be measured, whereas cortical tissue with type A pattern increased relatively (table II). This heterogeneity of the cortical epithelium could be identified more easily in stainings for HLA-Dr rather than for HLA-A,B,C. Thus, the quantitative estimations given in table II are lower for HLA-A,B,C in comparison to HLA-Dr.

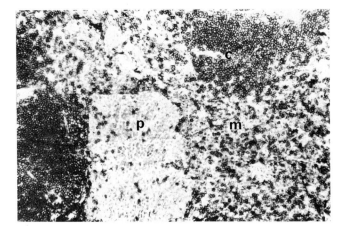

Fig. 4. Thymus, 2 years, male. Immunoperoxidase, anti-Leu-6. Note a large perivascular space (p) in the center. Many positive lymphocytes, some in small groups, as well as IDCs (large cells with positive peripheral reaction) can be seen within the medulla (m). c = Cortex. Counterstained. × 70.

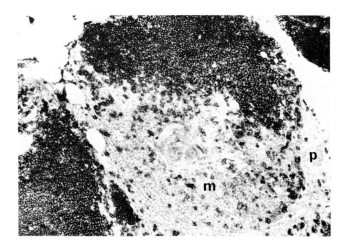

Fig. 5. Thymus, 43 years, male. Immunoperoxidase, anti-Leu-6. Decreased number of positive lymphocytes and IDCs within the medulla in comparison to figure 4. m = Medulla; p = perivascular space. Counterstained. × 70.

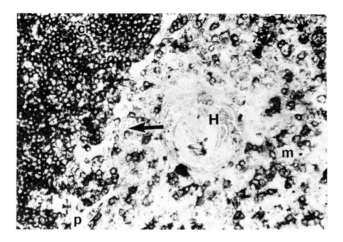

Fig. 6. Thymus, 11 years, female. Immunoperoxidase, anti-Leu-2a. c = Cortex; arrow = corticomedullary region; m = medulla; H = Hassall's corpuscle; p = perivascular space. Counterstained. ×280.

Epithelial cells and lymphocytes of the corticomedullary region and the medulla, interdigitating reticulum cells (IDC), and Hassall's corpuscles stained intensely for both antigens. A confluent pattern (type C) was expressed (fig. 3). During aging no change was observed. Thymic nurse cells could not be identified using monoclonal antibodies to HLA-Dr or HLA-A,B,C. Perivascular spaces increased quantitatively with age and exhibited a more or less depleted staining pattern for both antigens, HLA-Dr and HLA-A,B,C.

T Cell Differentiation Antigens

Results of staining for differentiation antigens of the Leu series is given in table II. Within the cortex of the thymus all antigens could be found by an intense peripheral staining of lymphocytes, except Leu-1. During aging, the intensity and the quantity of staining of these antigens did not change. Remnants of cortical tissue with unchanged presence of Leu-6, Leu-2a, Leu-3a, and Leu-4 could be identified in biopsies from patients with ages up to the sixth decade (fig. 4–8). Leu-1 exhibited a very low presence on cortical lymphocytes in biopsies of young persons. With increasing age, results of staining for Leu-1 increased as well. In the medulla, intensity and quantity of staining for Leu-1, Leu-4 and Leu-3a did not alter with age. However, numbers of Leu-2a-positive cells increased, whereas numbers of Leu-6-posi-

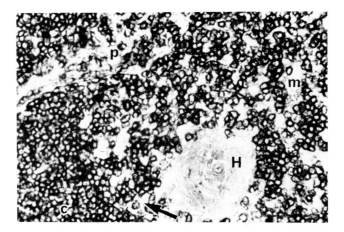

Fig. 7. Thymus, 20 years, female. Immunoperoxidase, anti-Leu-3a. c = Cortex; arrow = corticomedullary region; m = medulla; H = Hassall's corpuscle; p = perivascular space. Counterstained. ×280.

Fig. 8. Thymus, 68 years, male. Immunoperoxidase, anti-Leu-4. Small island of thymic lymphoid tissue with a cortical immunohistologic staining pattern. Counterstained. ×70.

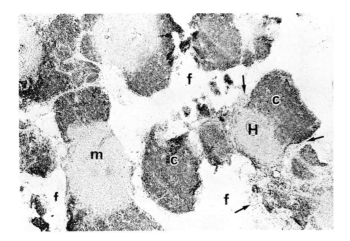

Fig. 9. Thymus, 42 years, male. Immunoperoxidase, anti-TdT. A thymic biopsy with little involution shows an intact distribution of TdT-positive cortical lymphocytes. c = Cortex; m = medulla; H = Hassall's corpuscle; f = fat tissue; arrow = perivascular lymphatic tissue. No counterstain. × 17.

tive cells decreased. Leu-6-positive IDCs could be distinguished based on their size. Especially in biopsies of young persons, Leu-6-positive island of cortical lymphocytes could be found within the medulla close to the corticomedullary region (fig. 4). Finally, the perivascular space contained more Leu-2a-positive cells in biopsies of persons with higher ages (table II).

Terminal Transferase

Immunohistologic determinants of TdT could be found related to the nucleus of most of all cortical lymphocytes. The borderline between cortex and medulla followed a distinct interdigitating course. Additionally, there were – comparable to the staining of Leu-6 – small TdT-positive groups of cells in the medulla close to the corticomedullary region. TdT-positive cortical lymphocytes could be found in all biopsies of patients with ages up to the sixth decade (fig. 9).

Discussion

In our study, the lymphoid cell population within the thymus has been identified according to their surface differentiation antigens. We obtained the information that there is obviously no major change of the differentiation pathway of T cell precursors in the human thymus during aging. Leu-1, a pan-T cell marker present on all circulating thymus-derived lymphocytes [5], was constantly found on both cortical and medullary lymphocytes throughout life. Its increasing presence on cortical lymphocytes may suggest a maturation effect. The increase of numbers of medullary and perivascular Leu-2a-positive cells, which represent cytotoxic/suppressor T subsets [6], is paralleled by findings in peripheral blood [9, 18]. Leu-6, an early differentiation antigen restricted to the thymus cortex, was found in biopsies staining from patients of all ages. That indicates that immature T cell precursors are present in the thymus throughout the end of human life.

This hypothesis is supported by findings using antibodies directed against terminal transferase. The fate of this unique DNA-polymerase during aging of the human thymus was unknown. Preliminary data obtained through biochemical determinations suggested a rapid disappearance of TdT from the human thymus after puberty [19]. Immunohistologically demonstrated TdT determinants in thymus biopsies of old persons of our study, on the contrary, indicate that the thymus of adults and seniors may continue to preserve a microenvironment which is either able to attract TdT-positive T cell-precursors from bone marrow [22] or spleen [1] or is able to induce synthesis of TdT by producing thymosin [20] or thymopoietin [12] in the epithelial cells.

Taken together, findings of differentiation antigens and terminal transferase suggest that the human thymus continuously maintains T cell differentiation during aging, and provides newly generated T cell clones throughout life.

Immunohistologic studies of the epithelial cell of the murine thymus of *Van Ewijk* et al. [7] have recently suggested to distinguish heterogenous thymic microenvironments based on the expression of MHC-coded antigens of epithelial-reticular cells and their processes. Our present investigation of the human thymus and its changes during aging has followed this understanding and led to the recognition of a heterogeneity not only of epithelial cells of the cortex, the medulla and the perivascular space, but also of a heterogeneity of epithelial cells within the cortex itself. Two distinct staining patterns (type A: reticular pattern, and type B: depleted pattern with faint peripheral positive reactions of lymphocytes) characterize different cortical epithelial microenvironments. Type A pattern is a tender reticular network straightened with an axis perpendicular to the thymus capsule and the corticomedullary borderline. This pattern of epithelial cells was described in the literature before [2, 12] as was the type C pattern (i.e. confluent pattern of the medulla). Type B pattern was observed before in certain areas of the medulla of murine thymuses and compared with that of the perivascular spaces [7].

Cortical tissue compartments that exhibit type B patterns of epithelial determinants in immunoperoxidase stainings, do not contain HLA-Dr-positive thymic nurse cells and correspond to areas free of epithelial cells as seen under the electron microscope [unpubl. observations]. Stereologically, type B pattern implies a high number of undifferentiated T cell precursors in the cortex without contact to the educating epithelial cells. The nature of these areas is presently unknown. However the striking similarity of their epithelial pattern with that of the perivascular spaces, which are forerunners of the fat atrophy of the thymus, may support the hypothesis that these cortical areas exhibit first signs of a degenerative process by a progressive loss of epithelial structures. This might be caused by either (1) an intrinsic functional decline of epithelial cells as non proliferating cells, or (2) a sequestration of certain cortical regions induced by physiological thymic functions. The latter alternative would be supported by the finding of a regional dependency of heterogeneity of the cortical thymic microenvironment.

References

1 Basch, R.S.; Kadish, J.L.; Goldstein, G.: Hematopoietic thymocyte precursors. IV. Enrichment of the precursors and evidence for heterogeneity. J. exp. Med. *147:* 1843–1848 (1978).
2 Bhan, A.K.; Reinherz, E.L.; Poppema, S.; McCluskey, R.T.; Schlossmann, S.F.: Location of T cell and major histocompatibility complex antigens in the human thymus. J. exp. Med. *152:* 771–782 (1980).
3 Bollum, F.J.: Antibody to terminal deoxynucleotidyl transferase. Proc. natn. Acad. Sci. USA *72:* 4119–4122 (1975).
4 Cantor, H.; Weissman, I.L.: Development and function of subpopulations of thymocytes and T lymphocytes. Prog. Allergy, vol. 20, pp. 1–64 (Karger, Basel 1976).
5 Englemann, E.G.; Warnke, R.; Fox, R.I.; Levy, R.: Studies of a human T lymphocyte antigen recognized by a monoclonal antibody. Proc. natn. Acad. Sci. USA *78:* 1791–1795 (1981).
6 Evans, R.L.; Wall, D.W.; Platsoucas, C.D.; Siegal, F.P.; Fikrig, S.M.; Testa, C.M.; Good, R.A.: Thymus-dependent membrane antigens in man: inhibition of cell-mediated lympholysis by monoclonal antibodies to the TH_2 antigen. Proc. natn. Acad. Sci. USA *78:* 544–548 (1981).
7 Ewijk, W. van; Rouse, R.V.; Weissman, I.L.: Distribution of H-2 microenvironments in the mouse thymus. Immunoelectron microscopic identification of I-A and H-2K bearing cells. J. Histochem. Cytochem. *28:* 1089–1099 (1980).

8   Gaudecker, B. v.; Müller-Hermelink, H.K.: Ontogeny and organization of the stationary non-lymphoid cells in the human thymus. Cell Tissue Res. *207:* 287–306 (1980).
9   Gupta, S.; Good, R.A.: Subpopulations of human T lymphocytes. X. Alterations in T, B, third population cells and T cells with receptors for immunoglobulin M (Tμ) or G (Tγ) in aging humans. J. Immun. *122:* 1214 (1979).
10 Hirokawa, K.: The thymus and aging; in Makinodan, Yunis, Immunology and aging, pp. 51–72 (Plenum Press, New York 1977).
11 Janossy, G.; Thomas, J.A., Eden, O.B.; Bollum, F.J.: Nuclear terminal deoxynucleotidyl transferase (TdT) in leukaemic infiltrates of testicular tissue. EMBO Wkshop on 'Terminal transferase – Immunbiology and leukemia' Elba 1981.
12 Janossy, G.; Thomas, J.A.; Bollum, F.J.; Granger, S.; Pizzolo, G.; Bradstock, K.F.; Wong, L.; McMichael, A.; Ganeshaguru, K.; Hoffbrand, A.V.: The human thymic microenvironment. An immunohistologic study. J. Immun. *125:* 202–212 (1980).
13 Kay, M.M.B.: The thymus. Clock for immunologic aging? J. invest. Derm. *73:* 29–38 (1979).
14 Kay, M.M.B.; Makinodan, T.: Relationship between aging and the immune system. Prog. Allergy, vol. 29, pp. 134–181 (Karger, Basel 1981).
15 McMichael, A.: HLA restriction of human cytotoxic T lymphocytes specific for influenza virus. J. exp. Med. *48:* 1458–1465 (1978).
16 Mertelsmann, R.: Leukämien und maligne Lymphome: Phänotypische und pathophysiologische Untersuchungen und ihre klinische Bedeutung (Thieme, Stuttgart 1981).
17 Modak, M.J.; Mertelsmann, R.; Koziner, B.; Pahwa, R.; Moore, M.A.S.; Clarkson, B.D.; Good, R.A.: A micromethod for determination of terminal deoxynucleotidyl transferase (TdT) in the diagnostic evaluation of acute leukemias. Cancer Res. clin. Oncol. *98:* 91–98 (1981).
18 Onsrud, M.: Age dependent changes in some human lymphocyte sub-populations: Changes in natural killer cell activity. Acta pathol. microbiol. scand, C, Immunol. *89:* 55–62 (1981).
19 Pahwa, R.N.; Modak, M.J.; McMorrow, T.; Pahwa, S.; Fernandes, G.; Good, R.A.: Terminal deoxynucleotidyl transferase (TdT) enzyme in thymus and bone marrow. I. Age associated decline of TdT in humans and mice. Cell. Immunol. *58:* 39–48 (1981).
20 Pazmino, N.H.; Ihle, J.N.; Goldstein, A.L.: Induction in vivo and in vitro of terminal deoxynucleotidyl transferase by thymosin in bone marrow cells from athymic mice. J. exp. Med. *147:* 708–718 (1978).
21 Reinherz, E.L.; Schlossman, S.F.: The differentiation and function of human T lymphocytes. Cell *19:* 821–827 (1980).
22 Silverstone, A.E.; Cantor, H.; Goldstein, G.; Baltimore, D.: Terminal deoxynucleotidyl transferase is found in prothymocytes. J. exp. Med. *144:* 543–548 (1976).
23 Sprent, J.: Antigen-induced selective sequestration of T-lymphocytes: role of the major histocompatibility complex; in Trnka, Cahill, Monogr. Allergy, vol. 16, pp. 233–244 (Karger, Basel 1980).
24 Stein, H.; Gerdes, J.; Schwab, U.; Lemke, H.; Mason, D.Y.; Ziegler, A.; Schienle, W.; Diehl, V.: Identification of Hodgkin and Sternberg-Reed cells as a unique cell type derived from a newly-detected small-cell population. Int. J. Cancer *30:* 445–459 (1982).
25 Weissman, I.L.; Rouse, R.V.; Kyewski, B.A.; Lepaut, F.; Butcher, E.C.; Kaplan, H.S.; Scollay, R.G.: Thymic lymphocyte maturation in the thymic microenvironment. Behring Inst. Mitt. *70:* 242–251 (1982).

26 Wekerle, H.; Ketelsen, V.P.; Ernst, M.: Thymic nurse cells. Lymphoepithelial cell complexes in murine thymuses: morphological and serological characterization. J. exp. Med. *151:* 925–944 (1980).

27 Yunis, E.J.; Fernandes, G.; Good, R.A.: Aging and involution of the immunological apparatus; in Twomey, Good, The immunopathology of lymphoreticular neoplasms, pp. 53–80 (Plenum Press, New York 1978).

28 Zinkernagel, R.M.; Callaham, G.N.; Althage, A.; Cooper, S.; Klein, P.A.; Klein, J.: On the thymus in the differentiation of 'H-2 self-recognition' by T cells. Evidence for dual recognition? J. exp. Med. *147:* 882 (1978).

Dr. G.G. Steinmann, Pathologisches Institut der Universität Kiel, Hospitalstrasse 42, D-2300 Kiel (FRG)

IV. Gene Expression and Nuclear Ageing

Molecular Mechanisms Decisive for Neuronal Ageing: A New Theory on Senescent Cellular Deterioration

H.-I. Sarkander

Neuropathologisches Institut der Freien Universität, Universitätsklinikum Steglitz, Berlin

> 'A theory is the more impressive the greater is the simplicity of its premises, the more different are the kinds of things it relates and the more extended is its range of applicability.'
> *Albert Einstein*

All present theories on ageing have one view in common: that the biological parameters varying as a function of ageing reflect modifications in the expression of genetic information, but no clear molecular mechanism has as yet been established. Expression of genetic information lastly means the realization of the information encoded in DNA in the form of functionable proteins warranting cellular survival. A large number of biochemical and molecular events are involved in this complex process at different cellular levels (table I). Therefore the general assumption was that any disturbance of one of these events could be the cause of cellular and organismal ageing as a result of progressive accumulation of errors in the genetic information system. Reflections of such nature and the possibility of detecting changes in the genetic information and/or in the transfer of genetic information at the different cellular levels brought about the development of a large number of theories on ageing. However, it should be stressed that these theories do not really represent independent theories of ageing but merely describe different aspects of age-dependent modifications in the expression of genetic information without distinguishing between specific and unspecific processes causing or accompanying ageing. Moreover, these theories do not take into account that a change in the mode and temporal sequence of gene readout per se already causes detectable modifications of genetic information at the following levels of gene expression without any primary disturbance having occurred at these levels.

Table I. Levels at which age-dependent modifications in gene expression may occur

Genome
  DNA
    Transcription
    Replication
    DNA repair
    Modification of DNA base pairs

  Chromosomal proteins
    Postsynthetic modification
    Interhistone-histone-DNA interactions
    Inter-NHC protein-NHC protein-DNA interactions

Nucleoplasm
  RNA processing
  Protein modification

Cytoplasm
  RNA processing
  Protein synthesis
  Protein processing

In fact these theories should be summarized to form one theory of genetic alteration in old age. To justify the unification of these different theories evidence must be furnished that the alteration of genetic expression at the different cellular levels in the last instance is due to one common causal basic mechanism. From a theoretical point of view a fundamental mechanism causing ageing is much more probable than the theory that the species-specific process of ageing is brought about in every individual by individually different processes. Moreover it is probable that increasing dysregulation of mechanisms, normally involved in gene expression, are the cause of the triggering and progress of ageing and that these are mechanisms efficient in the control and regulation of initial steps of gene expression, i.e. transcription of genetically active DNA sequences within the genome.

The initial and one of the decisive control mechanisms of gene expression is the transcription process. This process is composed of several highly specific steps, in particular the enzymatic transfer of information from DNA to RNA, catalyzed by the nuclear RNA polymerases, and the specificity with which selected DNA sequences are transcribed at certain times with a certain intensity. Due to the fact that the genetic information is not expressed simultaneously but sequentially, cellular events are attributed a temporal sequence and an irreversibility: every cellular event is characterized by its relative position with respect to the preceding step and by the temporal difference between the latter and previous events. This indicates that DNA not only serves as a carrier of genetic information, but also as a matrix programming the temporal sequence of events to be realized for guaranteeing cellular and at least organismal survival. Considerations of this kind are of special importance for evaluating the molecular mechanisms involved in the control and regulation of the sequential transcription of genetic information. It seems much more likely that dysregulation of mechanisms involved in sequential transcription is causally implicated in senescent deterioration of cellular functions than changes in the genetic information itself, or changes in the fidelity of readout of genetic information, and/or changes in the processing or transfer of genetic information, though they may also occur.

All experiments concerning age-dependent compositional changes of chromatin and chromatin-templated transcription performed to date have been carried out using total chromatin. Since, however, in differentiated cells only a small portion of selected DNA sequences ($< 10\%$) is transcribed into RNA, results obtained from investigations carried out on total chromatin are related to chromatin mainly in its transcriptionally inactive, biologically repressed state. For the purpose of recognizing regulatory mechanisms and components involved in the sequential transcription and in the age-specific modifications of the transcription process, it becomes necessary to fractionate chromatin into its transcriptionally active and inactive DNA sequences. This makes it possible to examine the compositional and functional properties of those DNA sequences that in vivo code for the cellular RNA. Moreover, since recent results employing deoxyribonuclease (DNase) have shown that the structure of nucleosomes is different in transcriptionally active and repressed DNA sequences, fractionation of chromatin in its template-active and inactive portions becomes indispensable in order to be able to recognize molecular mechanisms of regulation of gene expression at the transcriptional level.

To date, the method of choice for chromatin fractionation which yields a chromatin fraction enriched in its content of DNA sequences complementary to cellular RNA is the DNase II-Mg^{2+}-solubility method [9, 19], involving selective cleavage of chromatin with DNase II, followed by fractionation of the released portion on the basis of its solubility properties in 2 mM $MgCl_2$.

The present study contains a description of the compositional and functional age-dependent changes of DNase II-fractionated neuronal rat brain chromatin that gave rise to the concept of one basic theory of ageing, i.e. the theory of dysregulation of the sequential transcription of genetic information.

Results

Age-Dependent Compositional Properties of DNase II-Fractionated Neuronal Chromatin

Rapid and reliable fractionation of neuronal cerebral rat brain chromatin in transcriptionally active and repressed portions was achieved employing the DNase II-Mg^{2+}-solubility method [9, 19]. By comparison with the mechanical fractionation procedure, the nucleolytic digestion of chromatin has the advantage of avoiding deterioration of the chromatin substructure by pull-off histones or by sliding of histones along the nucleosomal DNA [4]. The differences in the Mg^{2+} solubility of the fragments obtained by nucleolytic cleavage leads to a preferential enrichment of non-repetitive unique DNA sequences in the Mg^{2+}-soluble transcriptionally active fraction, while the Mg^{2+}-insoluble fraction is enriched by highly repetitive repressed DNA sequences as revealed by RNA excess hybridization with labeled non-repetitive DNA as tracer [8].

Table II shows the quantities of transcriptionally active and repressed portions of neuronal chromatin and the age-dependent shift of this relation. The fact that neuronal chromatin isolated from adult animals has a higher content of transcribable DNA sequences than the neuronal chromatin of senile animals is clearly visible. This proves that in the course of ageing structural modifications of neuronal chromatin develop which age-dependently restrict the transcription of transcribable DNA sequences originally available in neurons of adult animals. This age-dependent increase of genetic restriction of neuronal chromatin agrees with the age-dependent reduction of the number of neuronal RNA species [25].

In order to understand better the role chromosomal proteins may play for genetic restriction and transcription of neuronal chromatin and its age-dependent modifications, as well as for the different structure of nucleosomes in transcriptionally active and repressed chromatin [6, 28], histones and non-histone chromosomal (NHC) proteins were comparatively determined quantitatively and qualitatively in template active and inactive neuronal fractions derived from adult and senile rats.

As summarized in table III the amount of NHC proteins on either a DNA or histone basis is significantly greater in the template active than in

Table II. Portion expressed in percent of chromosomal DNA of DNase II-fractionated neuronal chromatin

Chromatin	Age of animals months	DNase II-sensitive	Template active	Template inactive
N	12	87.4 ± 5.2	20.8 ± 2.1	66.1 ± 3.7
N	30	89.5 ± 5.7	13.3 ± 1.9	75.6 ± 5.2

Neuronal (N) chromatin was isolated from brains of 12-month-old and more than 30-month-old rats and suspended in 25 mM sodium acetate, pH 6.6, and treated with DNase II (8 enzyme units per A_{260} unit of chromatin for 15 min at 25 °C). The reaction was terminated by raising the pH to 7.5 with 0.1 M Tris-HCl, pH 10. The DNase II-resistant chromatin fraction was removed by centrifugation at 25,000 g for 20 min and the supernatant containing DNase II-sensitive chromatin is fractionated into transcriptionally active and inactive portions by addition of $MgCl_2$ up to a final concentration of 2.0 mM, then separated by centrifugation (25,000 g, 20 min). The DNA content of the different fractions was estimated by absorbance at 260 nm in 1 N NaOH (1.0 A_{260} = 38 μg/ml). Values given are arithmetic mean values ± SD of 5 preparations.

Table III. Chemical composition of DNase II-fractionated neuronal chromatin fractions

chromatin fraction	% DNA of unfractionated chromatin	n	Histone: DNA[1]	NHC protein: DNA[1]	RNA: DNA
N_{12} template active	20.8	6	0.82 ± 0.04	1.29 ± 0.05	0.015
N_{12} template inactive	66.1	6	1.07 ± 0.07	0.62 ± 0.03	0.010
N_{30} template active	13.3	4	0.79 ± 0.04	0.87 ± 0.045	0.027
N_{30} template inactive	75.6	4	1.09 ± 0.08	0.51 ± 0.031	0.0099

The fractions were isolated from 12-month-old (N_{12}) and more than 30-month-old rats (N_{30}). Chromatin isolation, DNase II fractionation, and the determination of the amounts of DNA, RNA, histones and non-histone chromosomal (NHC) proteins were carried out as reported elsewhere [19].
n = Number of determinations.
[1] Arithmetic mean value ± semi-error of the middle.

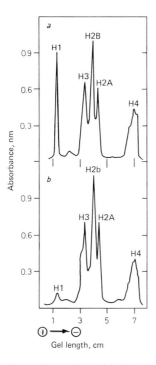

Fig. 1. Comparison of the densitometric scans of urea-polyacrylamide gel electrophoretically separated histone fractions of transcriptionally inactive *(a)* and transcriptionally active *(b)* neuronal chromatin fractions of adult rats. 10 OD_{260} units of neuronal chromatin were nucleolytically fractionated into transcribable and repressed portions by DNase II (8 units DNase II/A_{260} unit of chromatin) as described elsewhere [8]. Histones of each chromatin fraction were extracted with H_2SO_4 (0.4 N final concentration) and electrophoretically analyzed on polyacrylamide gels [8, 19].

the corresponding repressed fractions. The quantitative comparison of NHC protein content between the transcriptionally active and between the repressed fractions shows an age-dependent decrease of NHC proteins which is most marked in the genetically active fractions. On the contrary the histone-DNA mass ratios do not show any age-dependent modification though they are different between transcriptionally active and inactive fractions. Polyacrylamide gel electrophoresis of total histones revealed that the transcriptionally active neuronal fraction in both age groups is almost completely depleted of histone fraction H 1 (fig. 1). The importance of this finding results from the biological function the histone H 1 molecule has. It is assumed that the H 1 histone, which is associated with the linker DNA of nucleosomes, is

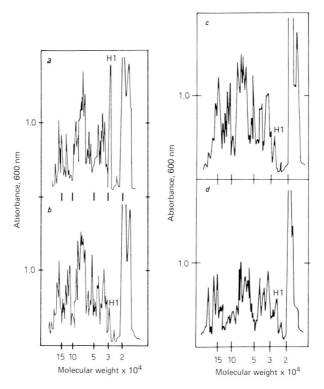

Fig. 2. a, b Comparison of the densitometric scans of SDS-polyacrylamide gel electrophoretically separated chromosomal proteins isolated from transcriptionally inactive *(a)* and transcriptionally active *(b)* neuronal chromatin of adult rats. Neuronal chromatin was fractionated into transcriptionally active and inactive fractions employing the DNase II/Mg^{2+}-solubility method [8, 19]. Total chromosomal proteins were extracted by Tris-SDS-β-mercaptoethanol buffer and applied to 10% SDS-acrylamide gels [19]. Gels run left to right, stained with Coomassie brilliant blue and scanned at 600 nm using a Gilford spectralphotometer with linear transport. *c, d* Comparison of densitometric scans of SDS-polyacrylamide gel electrophoretically separated, Coomassie blue stained chromosomal proteins isolated from transcriptionally active *(c, d)* neuronal chromatin fractions of adult *(c)* and senile *(d)* rats. Chromatin isolation and fractionation, protein extraction, and conditions of SDS-polyacrylamide gel electrophoresis were the same as described in the legend to figure 2a, b [10, 19].

involved in the generation and in the control of the maintenance of higher supranucleosomal chromatin structures [11, 24]. Moreover, the removal of the H 1 histone from the linker DNA is presumed to be essential for the accessibility and binding of regulatory proteins or enzymes to the linker DNA and for effective transcription [1, 3].

For assessing the obvious differences in the total amount of NHC proteins in the different fractions, NHC proteins were characterized on the basis of their molecular size by SDS-polyacrylamide gel electrophoresis. Figure 2a and b shows the densitometric absorption profiles of NHC proteins stained with Coomassie blue. Quantitative and qualitative differences between NHC proteins isolated from template active and repressed neuronal fractions are clearly visible. In particular, in addition to the decline of the NHC protein content, qualitative modifications of the NHC proteins occur in old age (fig. 2c, d).

The biological significance of these quantitative and qualitative differences in the composition of the NHC proteins in different chromatin fractions of neurons and the age-dependent modifications of the NHC protein composition is a consequence of the immanent function NHC proteins have in the regulation and control of gene expression. Although to date there is no clear evidence that NHC proteins can selectively influence the transcription of specific gene sequences, the idea that NHC proteins cause a modulation of the outflow of genetic information by means of their binding to DNA is generally accepted [5, 14]. Thus it is known that by binding NHC proteins to repetitive DNA sequences brings about an inhibitory effect on chromatin-templated transcription while other NHC proteins bind to unique DNA sequences causing activation of the readout of genetic information. In this connection I wish to draw attention to the importance of high mobility group (HMG) proteins belonging to the NHC proteins [7]. It is known that these proteins, especially HMG protein 14 and 17, preferably associate with specific nucleosomes localized in actively transcribed DNase I-sensitive gene sequences [29]. Moreover, it has been recognized that DNase I preferentially digests transcriptionally active DNA sequences containing, besides structural inactive nucleosomes, a small, up-to-date unknown number of active nucleosomes, characterized by their ability to specifically bind HMG proteins [28]. The selective removal of HMG proteins 14 and 17 from the nucleosomes of active gene sequences using 0.35 M NaCl leads to a loss of the DNase I sensitivity of these sequences [15] which can be reversed after addition of the HMG proteins [27, 30]. The added HMG proteins 14 and 17 will then specifically bind to the HMG-depleted active nucleosomes within the active chromatin, maximally 2 HMG protein molecules per nucleosome [30]. This shows that active nucleosomes have physicochemical properties making it possible for HMG proteins to bind specifically.

The attempt to determine comparatively the total number of active nucleosomes within template active neuronal chromatin fractions, derived

Table. IV. Quantitative determination of HMG 14 and 17 proteins associated with template active neuronal chromatin fractions prepared from 12-month-old (N_{12}) and over 30-month-old (N_{30}) rats

Chromatin fraction	µg HMG 14 and 17 proteins / mg chromosomal DNA
N_{12} template active	38.7 ± 2.6
N_{30} template active	51.2 ± 3.1

HMG 14 and 17 proteins were isolated and quantified by the method of *Goodwin and Johns* [7]. DNA was spectrophotometrically assayed by absorption at 260 nm.

from adult and senile animals by extracting the total amount of HMG proteins from the isolated template active neuronal fraction, fails because already during the fractioning of chromatin by DNase II the HMG proteins are partially removed from the chromatin, as can be seen from the HMG protein content of the supernatant of the DNase II digestion assays. Even when it was attempted to carry out the fractionation of chromatin with other DNA-digesting enzymes such as DNase I or micrococcal nuclease, it was not possible to avoid the partial removal of HMG proteins.

In the present study I determined the HMG protein-binding capacity of template active neuronal chromatin fractions by saturating the partially HMG-depleted chromatin fractions resulting from the DNase II fractionation with an excess of exogenously added HMG protein 14 and 17. Next I carried out repeated washings under low ionic strength conditions and removed the nonspecifically bound HMG proteins before extracting with 0.35 M NaCl the specifically nucleosomally bound part of the HMG proteins and quantifying them. The method chosen might have the disadvantage that the exogenously added HMG protein excess brings about nonspecific binding of HMG proteins to other chromosomal components and that even by repeated washing a complete removal of the nonspecifically nucleosomally bound HMG proteins cannot be achieved. However, measurements of the HMG protein content of partially HMG-depleted chromatin fractions showed that after different quantities of HMG proteins had been added to the assay, the protein content of the chromatin fractions remained constant independent of the HMG protein quantities added to the assays. This proves that there can be no concentration-dependent nonspecific binding of HMG

Table V. Transcription of template active neuronal chromatin fractions isolated from 12-month-old (N_{12}) and more than 30-month-old (N_{30}) rats

Chromatin fraction	UMP incorporation pmol
N_{12} template active	29.7 ± 1.9
N_{30} template active	55.6 ± 2.6

Template activity of the neuronal chromatin fractions (2.5 µg DNA) was assayed with 5 units of RNA polymerase B as detailed elsewhere [19] and expressed as picomoles of [^{3}H]-UMP incorporation per assay tube. Values given are means of quadricate determinations ± SD of 3 replicate experiments.

proteins to the chromatin or an insufficient removal of the excessive quantity of HMG proteins added.

Table IV shows the binding capacities for HMG proteins 14 and 17 of template active neuronal chromatin of different ages. In particular it can be seen that the HMG-binding capacity of the template active neuronal chromatin increases in old age. Since HMG proteins 14 and 17 bind preferentially to active nucleosomes within the chromatin, the age-dependent increase in the HMG-binding capacity of template active chromatin indicates that the number of active nucleosomes within the template active neuronal chromatin fraction is enriched.

In vitro Transcription of DNase II-Fractionated Neuronal Chromatin Isolated from Adult and Senile Rats

Nonrepetitive (unique) template active DNA sequences that have been proved to code for cellular RNA in vivo were enriched from neuronal cerebral chromatin isolated from the brain of rats, 12 months and older than 30 months, and used as templates for the RNA synthesis in vitro. As evidenced by table V, neuronal transcriptionally active chromatin isolated out of the total chromatin of senile rats is transcribed more efficiently than the corresponding neuronal template of adult rats when carried out under conditions allowing reinitiation of the exogenously added rat brain RNA polymerase B.

Several molecular events may be mechanistically involved in the increased transcription rate of neuronal chromatin in old age: (a) differences in the number of available RNA initiation sites; (b) differences in the initia-

Table VI. Acetylation of histones associated with template active neuronal chromatin fractions

Chromatin fraction	[^{3}H]-acetate incorporated dpm × μg histone^{-1}
N_{12} template active	1,743 ± 88
N_{30} template active	3,271 ± 147

Chromatin fractions were isolated from brains of 12-month-old (N_{12}) or over 30-month-old (N_{30}) rats. DNase II fractionation and acetylation of histones were the same as described previously [18, 22].

tion frequency with which the RNA polymerase-B molecules utilize the RNA initiation sites, and (c) different velocities of the RNA chain elongation. The share these different phenomena have in the age-dependent changes of neuronal transcription has been studied. The total number of RNA initiation sites available for brain RNA polymerase-B molecules was determined under conditions rendering reinitiation and unspecific initiation of RNA polymerase molecules absolutely impossible and warrant RNA polymerase saturation of the templates [2, 18, 23]. These comparative age-dependent measurements of the total number of RNA initiation sites on transcriptionally active neuronal chromatin fractions showed an identical number of RNA initiation sites on templates derived from 12- and 30-month-old rats. The identical number of available RNA initiation sites on the transcribable neuronal DNA sequences of different ages is remarkable particularly because I was able to show that the composition and structure of template active neuronal chromatin differ very considerably at both points in time [16, 23]. The different acetylation rates of histones in both chromatin fractions prove the difference in the structure of 'adult' and 'senile' template active chromatin. The fact that by comparison with the 'adult' fraction almost twice as much acetate is taken up by the core particle DNA-bound histones of the template active neuronal chromatin of senile rats clearly indicates that in the course of ageing modifications of histone binding occur within neuronal nucleosomes (table VI). To be able to judge the distribution of incorporated [^{3}H]-acetate between the different nucleosomal histone fractions, the acetylated neuronal histones prepared from DNase II-fractionated chromatin of both age groups were isolated and electrophoretically separated on urea-polyacrylamide gels (fig. 3) [13, 18].

Taking into account that histones are acetylized at their primary DNA binding sites, a higher acetate uptake of nucleosomal DNA-bound histones

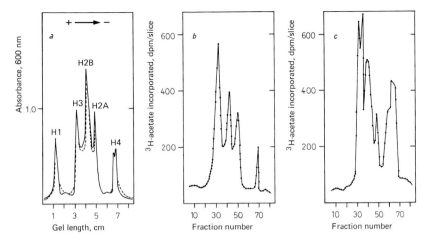

Fig. 3. Absorbance profiles and comparative electrophoretic analyses of the distribution of [^{3}H]-acetate uptaken by neuronal histones. *a* Densitometric tracings of the stained banding profiles of acetylated neuronal histones isolated from transcriptionally active chromatin of adult (——) or senile (······) rats are practically identical. *b* Distribution of [^{3}H]-acetate bound to neuronal histones of transcriptionally active chromatin derived from adult rats. *c* Distribution of [^{3}H]-acetate bound to neuronal histones of transcriptionally active chromatin derived from senile rats. Histones were isotopically labeled up to maximum values by incubating transcriptionally active neuronal chromatin fractions with saturating amounts of [^{3}H]-acetyl-coenzyme A. Aliquots of the assays were pipetted into ice-cold H_2SO_4 up to a final concentration of 0.4 N from which histones were extracted, purified, and subjected to polyacrylamide gel electrophoresis as described previously [18, 19].

signals a lower binding strength of histones to nucleosomal DNA [20]. Since histones, as their different acetylation rates in template active and inactive neuronal chromatin indicate [19], are less strongly bound in the template active than in the repressed DNA sequences, the additional increase of histone acetylation in old age shows a further loosening of histone-DNA binding in the template active sequence. From this the question arises, what is the biological and functional significance of the additional destabilization of histone-DNA interactions in the genetically active RNA-coding DNA sequences of the neuronal chromatin occurring in old age.

We have been able to show that the acetylation of chromatin brings about an increase in the number of available RNA initiation sites on the acetylated template and that the amount of acetate uptake correlates positively with the degree of increase of RNA initiation sites [17, 21]. On the basis of this functional relationship it is interesting to test whether the higher ace-

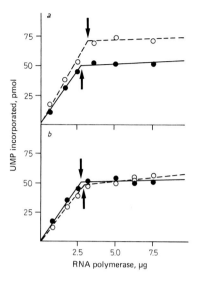

Fig. 4. Neuronal RNA polymerase B saturation curves of acetylated (○) and non-acetylated (●) transcriptionally active chromatin isolated from adult *(a)* and senile *(b)* rats. Nucleosomally DNA-bound neuronal histones were acetylated up to maximum values by incubating transcriptionally active neuronal chromatin with saturating amounts of acetyl-coenzyme A (120 μmol/μg DNA). For the quantitative determination of RNA initiation sites fixed amounts (5 μg DNA) of acetylated and non-acetylated transcriptionally active neuronal chromatin were titrated to saturation (arrows) with increasing amounts of RNA polymerase B. Isolation and standardization of RNA polymerase B and titration of the total number of RNA initiation sites were carried out as described [12, 18, 19].

tate uptake capacity of DNA-bound histones in old age is related to a higher availability of previously inaccessible RNA initiation sites. For this reason we determined by comparison the number of RNA initiation sites available for exogenously added RNA polymerase-B molecules on acetylated and non-acetylated template active neuronal chromatin fractions prepared from adult and senile rats. As can be seen from the different positions of the inflection points of the titration kinetics of RNA polymerase saturation curves of 'adult' neuronal chromatin (fig. 4a), the enzymatic acetylation of template active neuronal chromatin creates an additional availability of RNA initiation sites on the acetylated chromatin. The situation is different when a comparison is made between the number of RNA initiation sites of acetylated and non-acetylated neuronal template active chromatin isolated from senile rats. Here the almost identical position of inflection points (arrows in fig. 4b) of the titration curves indicates that acetylation of senile template active neuronal chromatin causes no or only a negligible increase in the number of RNA initiation sites on acetylated 'senile' chromatin. The fact that the acetylation effect does not appear is surprising. It could be interpreted as stating that the higher acetylation capacity of the histones localized in the template active sequences of the 'senile' chromatin by comparison with 'adult' chromatin may already be a compensatory mechanism which in old age allows maintenance of the number of RNA initiation sites in the order of magnitude measured in 'adult' chromatin. This would imply that the identical number (not identity) of RNA initiation sites on neuronal template active chromatin of adult and senile rats is caused and maintained by different molecular mechanisms, whereas the lack of appearance of an additional availability of RNA initiation sites by means of acetylation of 'senile' chromatin signifies that neuronal chromatin in old age is already transcribed with a capacity that cannot be increased further by the acetylation mechanism.

In summary it can be said that in old age the transcription of neuronal genetically active chromatin sequences has a capacity that can no longer be influenced in the sense of a further increase by such a decisive gene regulatory mechanism as histone acetylation. This means that, during the ageing of neurons, adaptative and regulative mechanisms of chromatin-templated transcription are lost or become ineffective while their effectiveness has been proved in neurons of adult animals. Besides this age-dependent reduced adaptability to varying needs of the transcription process, the number of RNA initiation sites on the template active chromatin of senile animals, which can no longer be increased by acetylation of DNA-bound histones, must be interpreted as evidence of an age-dependent increase of the degree of genetic restriction of neuronal chromatin. This increase of genetic restriction of neuronal chromatin in the course of aging grows to be even more impressive as a result of the age-dependent decrease in the number of transcribable DNA sequences, which becomes visible from the shift of the mass ratio between transcribable and repressed DNA sequences within neuronal chromatin in favor of the repressed portion (table II).

For further characterization of transcriptive properties of neuronal chromatin and its age-dependent modifications a comparative determination was made between the initiation frequency with which RNA polymerase-B molecules utilize the RNA initiation sites localized on template active DNA sequences of 'adult' and 'senile' neuronal chromatin and the elongation velocity with which neuronal RNA chains grow when synthesized by RNA polymerase-B molecules. The theoretical basis of the methods applied for

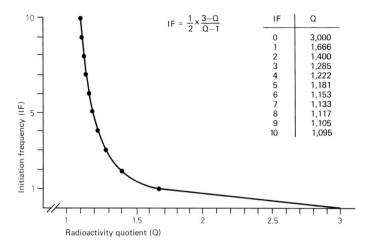

Fig. 5. Functional relationship between the number of initiations (IF) occurring during the period of incubation and the numerical value of the radioactivity quotient (Q) of double-labeled RNA. Assay composition and incubation conditions are described in the footnote to table VII. If RNA is synthesized under conditions allowing repeated initiations of RNA polymerase molecules at the same RNA initiation site, the sequential addition of isotopes, e.g. [^{14}C]-UTP and [α-^{32}P]-UTP, to the assay mixture at different stages of the completion of RNA chains results in differential incorporation of both isotopes into the RNA synthesized. The numerical value of the ^{14}C/^{32}P radioactivity ratio allows a calculation of the initiation frequency (IF).

determination of the initiation frequency and elongation velocity are described by several investigators [16, 19, 26] and depicted in figure 5. The different values of the [^{14}C]-UMP/[α-^{32}P]-UMP quotient of the double-labeled neuronal RNA indicate that RNA initiation sites localized on the template active DNA sequences of senile chromatin by comparison in time are more frequently utilized by RNA polymerase-B molecules than the RNA initiation sites available on the template active DNA sequences of adult chromatin. The efficiency with which RNA polymerase-B molecules incorporate nucleotides into the growing RNA chains is also higher in old age (table VII).

A final overall consideration of the results described permits the conclusion that the degree of genetic restriction of neuronal chromatin increases in old age, whereas the adaptability and regulability of the transcription process declines. Evidence justifying this conclusion can be seen in the age-dependent shift of the mass ratio of transcribable and repressed DNA sequences within neuronal chromatin and in the inability to make additional RNA initi-

Table VII. Age-dependent modifications of neuronal initiation frequency (IF) and elongation velocity (V_E) of neuronal RNA chain growth

Chromatin fraction	Q	IF	$\Delta IF_{N_{12}-N_{30}}$	V_E (nucleotides/s)	$\Delta V_{E\,N_{12}-N_{30}}$
N_{12} template active	1.1785	5.1		23	
			70.59		47.8
N_{30} template active	1.1086	8.7		34	

The functional connection between Q ($[^{14}C] - [\alpha-^{32}P]$-UMP quotient) and IF is represented in figure 5. Under assay conditions allowing repeated initiation of RNA polymerase B molecules at the same RNA initiation site, the initiation frequency (reinitiation efficiency) of rat brain RNA polymerase was measured as detailed [19, 26] by incorporation of $[^{14}C]$-UMP and $[\alpha - ^{32}P]$-UMP into newly synthesized neuronal RNA. Assays contained 2.5 µg chromatin template; 200 mM NH$_4$Cl, pH 7.9; 5 mM Mg-acetate; 2 mM ATP, GTP, and CTP; 0.08 mM of both $[^{14}C]$-UTP and $[\alpha - ^{32}P]$-UTP; 5 units RNA polymerase. The V_E of neuronal RNA chain growth was determined by the rate of incorporation of $[\gamma -^{32}P]$-ATP, $[\gamma -^{32}P]$-GTP, and $[^{14}C]$-UMP into the neuronal RNA synthesized under assay conditions completely preventing reinitiation [2, 18].

Table VIII. RNA content of neuronal nuclei isolated from brains of 12-month-old (N_{12}) and more than 30-month-old (N_{30}) rats (mean ± SD)

Cell nuclei	pg RNA/cell nuclei
N_{12}	0.12 ± 0.005
N_{30}	0.37 ± 0.007

Isolation of nuclei, RNA and DNA extraction, and RNA and DNA determination were carried out as reported previously [19].

ation sites available by acetylation of histones. As evidenced by the age-dependent increase of neuronal nuclear RNA (table VIII) the more frequent utilization of RNA initiation sites and the acceleration of the elongation velocity seem not to be compensatory mechanisms for allowing the synthesis of RNA quantities necessary for the survival of neurons but are more probably the expression of the loss of control and regulation mechanisms of transcription. In dependence of the degree of the loss of regulatory transcriptive mechanisms, the degree of orderly sequential transcription of genetic information may be disturbed leading to senescent deterioration and finally to a 'lethal imbalance' in the 'concerted actions' of neuronal metabolism.

Conclusions

This paper presents results of experiments carried out on isolated DNase II-fractionated (EC 3.1.4.6) neuronal cerebral chromatin of rats of different age. It is shown that: (1) the degree of genetic restriction of neuronal chromatin increases age-dependently as can be seen from the shift of the mass ratio of transcribable and repressed DNA sequences within neuronal chromatin; (2) the structure of the quantitatively reduced transcriptionally active DNA sequences of neuronal chromatin is age-dependently modified as evidenced by the higher capacity of acetylation of core particle DNA-bound histones and by the increased HMG 14 and 17 protein content of template active neuronal DNA sequences; (3) there is a functional connection between the age-dependent modification of the structure of transcriptionally active neuronal chromatin and the increase of the neuronal rate of transcription leading to an increased nuclear neuronal RNA content, and (4) the age-dependent increased transcription rate of genetically active neuronal chromatin can be traced back to more frequent utilization of RNA initiation sites localized on the transcriptionally active neuronal chromatin as well as to a higher elongation velocity of RNA chain growth catalyzed by RNA polymerases.

The increased transcribability of template active neuronal chromatin in old age visible from the increased initiation frequency and the rising elongation velocity can be attributed to two molecular mechanisms of which one is the increased HMG 14 and 17 protein content of transcribable DNA sequences, leading to an increased number of active nucleosomes within the transcribable DNA sequences, and the other the higher capacity of acetylation of nucleosomal core particle DNA-bound histones, indicating a loosening of histone binding to DNA, known to facilitate chromatin-templated transcription by reducing the steric hindrance of RNA polymerases. Since HMG proteins are known to inhibit histone deacetylases [29] and thereby to increase the acetate content of histones the increase of HMG 14 and 17 proteins in old age leads to an additional increase in the acetate content of

DNA-bound histones. The consequence of this is another extraordinary increase of template transcribability.

Since chromatin is not only carrier of genetic information but also a matrix of information for timing the readout of genetic information it is reasonable to assume that the increased, possibly maximum RNA synthesis in old age expresses a lack of regulability of the outflow of genetic information connected with a dysregulation of the sequential transcription and thereby with a disturbance of an orderly realization of genetic information. The accumulation of neuronal RNA transcriptates in the nucleus occurring in old age justifies the above assumption.

This RNA accumulation in neuronal nuclei may be the consequence of a temporal and capacitive dissociation of the molecular mechanisms involved in gene expression at different cellular levels. Temporal shifts in the order of transcription of selected DNA sequences result in a shift in time in the chains of molecular events which must be realized in order to guarantee cellular survival.

The theory on the dysregulation of sequential transcription of genetic information offers a satisfactory explanation for the different genetic aspects which form the basis for the concepts of a large number of theories on ageing and traces all hitherto known deteriorations of cellular functions occurring in old age back to one general disturbance – the dysregulation of sequential transcription.

References

1 Allan, J.; Hartmann, P.G.; Crane-Robinson, S.; Aviles, F.: The structure of histone H1 and its location in chromatin. Nature, Lond. *288:* 657–679 (1981).
2 Axel, R.; Cedar, H.; Felsenfeld, G.: Chromatin template activity and chromatin structure, Cold Spring Harb. Symp. quant. Biol., vol 38, pp. 773–783 (Cold Spring Harbor Laboratory, Cold Spring Harbor 1974).
3 Diez-Caballero, T.; Aviles, F.; Albert, A.: Specific interaction of histone H1 with eukaryotic DNA. Nucl. Acids Res. *9:* 1382–1392 (1981).
4 Doenecke, D.; McCarthy, B.J.: Movement of histones in chromatin induced by shearing. Eur. J. Biochem. *64:* 405–409 (1976).
5 Elgin, S.C.R.; Weintraub, H.: Chromosomal proteins and chromatin structure. Rev. Biochem. *44:* 725–774 (1975).
6 Garel, A.; Axel, R.: Selective digestion of transcriptionally active ovalbumin gene from oviduct nuclei. Proc. natn. Acad. Sci. USA *73:* 3966–3970 (1976).
7 Goodwin, G.H.; Johns, E.W.: Are the high mobility group non-histone chromosomal proteins associated with active chromatin? Biochim. biophys. Acta *519:* 279–284 (1978).

8   Gottesfeld, J.M.; Bagi, G.; Berg, B.; Bonner, J.: Sequence composition of the template-active fraction of rat liver chromatin. Biochemistry 15: 2472–2483 (1976).
9   Gottesfeld, J.M.; Garrad, W.T.; Bagi, G.; Wilson, F.R.; Bonner, J.: Partial purification of template-active fraction of chromatin. Proc. natn. Acad. Sci. USA 71: 2193–2197 (1974).
10 Gottesfeld, J.M.; Partington, G.A.: Distribution of messenger RNA coding sequence in fractionated chromatin. Cell 953: 962 (1977).
11 Lawrence, J.-J.; Goeltz, P.: Distribution of linker DNA in relation to core DNA in H1 depleted nucleosomes. Nucl. Acid Res. 9: 859–866 (1981).
12 Lux, R.M.; Cervos-Navarro, J.; Sarkander, H.-I.: Modifications in the organization of neuronal nucleosomes in the course of aging; in Cervos-Navarro, Sarkander, Brain aging: neuropathology and neuropharmacology, pp. 329–349 (Raven Press, New York 1983).
13 Panyim, S.; Chalkley, R.: High resolution acrylamide gel electrophoresis of histones. Archs Biochem. Biophys. 130: 337–346 (1969).
14 Saffer, J.D.; Coleman, J.E.: Reversible phosphorylation of nucleosome binding protein that stimulates transcription of nucleosome deoxyribonucleic acid. Biochemistry 19: 5874–5883 (1981).
15 Sandeen, G.; Wood, W.I.; Felsenfeld, G.: The interaction of high mobility proteins HMG 14 and 17 with nucleosomes. Nucl. Acids Res. 8: 3757–3778 (1980).
16 Sarkander, H.-I.: Age-dependent changes in the organization and regulation of transcriptionally active neuronal and non-astrocytic glial chromatin; in Cervos-Navarro, Sarkander, Brain aging: neuropathology and neuropharmacology, pp. 301–327 (Raven Press, New York 1983).
17 Sarkander, H.-I.; Brade, W.P.: Studies on the in vitro turnover of histone acetyl groups and the increase of RNA synthesis in rat liver nuclei after in vivo application of α-hexachlorocyclohexane. Biochem. Pharmac. 24: 2279–2285 (1975).
18 Sarkander, H.-I.; Dulce, H.-J.: Studies on the regulation of RNA synthesis in neuronal and glial nuclei isolated from rat brain. Exp. Brain Res. 31: 317–327 (1978).
19 Sarkander, H.-I.; Dulce, H.-J.: Characteristics of transcriptionally active and inactive neuronal and non-astrocytic glial rat brain chromatin fractions. Exp. Brain Res. 35: 109–125 (1979).
20 Sarkander, H.-I.; Fleischer-Lambropoulos, H.; Brade, W.P.: A comparative study of histone acetylation in neuronal and glial nuclei enriched rat brain fractions. FEBS Lett. 52: 40–43 (1975).
21 Sarkander, H.-I.; Knoll-Köhler, E.: Changing patterns of histone acetylation and RNA synthesis of the developing and ageing rat brain. FEBS Lett. 85: 301–304 (1978).
22 Sarkander, H.-I.; Knoll-Köhler, E.; Cervos-Navarro, J.: Repression of glial RNA transcription during the development of 6-aminonicotinamide (6-AN)-induced acute gliopathy. J. Pharmac. exp. Ther. 205: 503–514 (1978).
23 Sarkander, H.-I.; Lux, R.; Cervos-Navarro, J.: Histone-DNA interactions in neuronal chromatin during aging. Exp. Brain Res. 5: 45–50 (1983).
24 Smith, B.J.; Johns, E.W.: Histone H1: its location in chromatin. Nucl. Acids Res. 8: 6069–6079 (1981).
25 Strehler, B.L.: Genetic and neural aspects of redundancy and aging; in Schmidt et al., 5th European Symposium on Basic Research in Gerontology, pp. 36–61 (Perimed, Erlangen 1977).
26 Udvardy, A.; Seifart, K.: Transcription of specific genes in isolated nuclei from HeLa cells in vitro. Eur. J. Biochem. 62: 353–363 (1976).

27 Vidali, G.; Boffa, L.C.; Allfrey, V.G.: Selective release of chromosomal proteins during limited DNase 1 digestion of avian erythrocyte chromatin. Cell *12:* 409–415 (1977).
28 Weintraub, H.; Groudine, M.: Chromosomal subunits in active genes have an altered conformation. Science *193:* 848–856 (1976).
29 Weisbrod, S.; Groudine, M.; Weintraub, H.: Interaction of HMG 14 and 17 with actively transcribed genes. Cell *19:* 289–301 (1980).
30 Woodcock, C.L.F.; Frado, L.-L.Y.; Wall, J.S.: Composition of native and reconstituted chromatin particles. Proc. natn. Acad. Sci. USA *77:* 4818–4822 (1980).

Priv.-Doz. Dr. med. H.-I. Sarkander, Neuropathologisches Institut der
Freien Universität, Universitätsklinikum Steglitz, Hindenburgdamm 30,
D–1000 Berlin 45

Cellular Aging, Neoplastic Transformation, Meiotic Rejuvenation, and the Structure of Chromatin Complex

Sinan Taş

Tübitak Research Institute for Basic Sciences, Gebze, Kocaeli, Turkey

Aging, an Overview

Although organismal aging is a complex phenomenon influenced by supracellular organizational features, the basic aging process appears to derive primarily from the intrinsic failures of cells. This is attested by the aging of unicellular eukaryotes [1] and of cultured normal diploid cells of multicellular organisms [2]. Further, the number of population doublings preceeding the onset of senescence or 'crisis' in cultures of normal diploid cells has been shown to correlate with the age and maximum lifespan potential of donor organisms [2, 3].

Towards an understanding of the mechanisms of cellular aging, it seems useful to look at the phenomenon from evolutionary and comparative points of view. The answer to the question whether or not there are cell types that are potentially immortal is affirmative. Prokaryotic cells under optimal environmental conditions can propagate indefinitely. Similarly, even though individuals of probably all multicellular eukaryotes have limited lifespans, germ line cells of these organisms are capable of giving rise to youthful organisms at advanced ages. Also neoplastically transformed cell populations can be propagated indefinitely both in vitro and in vivo but normal cell populations ultimately die out under identical conditions [2, 4]. Examples of nontumorigenic cell lines that do not cease proliferating in culture are known, but these cells appear different than the early generation cells ('precrisis cells') growing from the primary explants [5].

What would be the common denominators for nonaging cell types? *Martin* [6], in a review of rejuvenation associated with meiosis, points out that meiosis is correlated with dramatically increased DNA repair activities, and suggests that immortality of germ lines and of sexually reproducing unicellular eukaryotes may be due to correction of DNA damages in meiosis. Another thesis dealing with cellular immortality is that cell aging may in large part be a consequence of the developmental and stochastic alterations of the chromatin complex that in turn alter transcriptive, replicative, as well as reparative functions of DNA, and that the nonaging cell types may avoid or repair aging changes of chromatin complex in one way or another [7–9]. Mounting evidence indicates that aging is indeed associated with a generalized impairment of DNA functions, whether replicative [10, 11], transcriptive [12] or reparative [13, 14].

Structure of Chromatin Complex in Normal Aging Cells and in Neoplastic Cells

Unlike the situation in prokaryotic cells, DNA of eukaryotic cells is extensively complexed with specialized proteins to form the chromatin complex in which accessibility of DNA is restricted. The first level of packaging of eukaryotic DNA, complexing with histones to form the nucleosomes, is sufficiently understood but relatively little is known about the supranucleosomal levels of organization. Nevertheless the work of a number of authors contributes to an emerging picture of chromatin structure, briefly summarized here.

When cell nuclei are demembranized by nonionic detergents and digested extensively with deoxyribonucleases, a skeleton structure of nuclear proteins, variously called nuclear matrix [15] or scaffold [16], is seen to remain and preserve the shape of nuclei. When intact cells are lysed with nonionic detergents and the released nuclear structures are treated with 2.0 M NaCl, most of the nonhistone proteins and histones dissociate from DNA to leave behind a subgroup of nuclear proteins tightly associated with DNA [17]. DNA of these 2.0 M NaCl-treated nuclear structures (called here DPC 2.0s) exists as negatively supercoiled loops, and their proteins appear to include most or all of the proteins of nuclear protein skeleton structures obtained by the deoxyribonucleolytic digestion of nuclei [16, 18, 19]. The significance of the association with DNA of these 2.0 M NaCl-resistant proteins in the regulation and mechanics of DNA replication and transcription is stressed by several recent findings [20–25].

Our studies of the chemical nature of interactions between the proteins of nuclear skeleton structures, and between DNA and these proteins indicated that extensive intermolecular disulfide (S-S) bonds link proteins of nuclear protein skeleton to one another as well as to regions of DNA (or to proteins or other molecules bound to DNA by covalent-like bonds), and that S-S bonds are involved in maintaining a condensed chromatin structure [7–9, 19, 26–30]. Recently *Kaufmann* et al. [31] also reported extensive S-S linkages between the proteins of nuclear matrices prepared from normal liver cells. In line with these findings [7–9, 19, 26–31] another recent study indicated that certain of the nonhistone proteins tightly associated with nuclear DNA can be dissociated by S-S-reducing agents [18].

With regard to aging, an earlier study revealed that chromatin material solubilized from rat liver nuclei and from nuclei of whole insects display greater S-S bonding with increasing age [7]. Later studies indicated that treatment of demembranized cell nuclei with S-S-reducing agents decrease their sedimentation rates through neutral sucrose gradients [9, 26, 27]. The decrease of sedimentation rates appeared to be due to decondensation of nuclei [9, 27] and happened to a greater extent with older animals [9, 26, 27]. Nuclear fractionation studies suggested that the relatively nuclease-resistant fraction, which presumably contains the nuclear protein skeleton or matrix, is the principal site of increased S-S bonding during aging [9, 26]. Other studies also pointed to an involvement of this nuclear fraction in the increase of S-S-mediated condensation of chromatin complex during aging [28]. Thus we found that, when a limited number of nicks are introduced to the DNA of demembranized nuclei by endogenous nucleases or by DNAse I under conditions that generate practically no acid soluble DNA, up to 80% of the DNA can be dissociated from the nuclear protein skeleton as soluble chromatin particles by hypotonic lysis of nuclei [28]. Additional DNA could be released from the nuclear protein skeletons by treatment with S-S-reducing agents, and to a significantly greater extent with liver cell nuclei from old than young mice [28]. *Ronne and Wulf* [32] found evidence that a condensation of chromatin may also occur during in vitro aging of cells.

In neoplastic cell populations reversion to aging and terminal differentiation can occur in a fraction of clones [33], but evidently at a low enough frequency under ordinary conditions to allow for the unlimited propagation. To investigate the S-S-mediated condensation of the nuclear DNA-protein complexes (DPC's) of neoplastic cells, we developed techniques that enabled us to assess the S-S-mediated DPC condensation in individual cells. For this purpose living cells were lysed with a nonionic detergent in the presence of various concentrations of NaCl plus a sulfhydryl (SH) blocking agent which prevents SH oxidation and SH/S-S interchange reactions [30]. The nuclear

Table I. Influence of treatment with S-S-reducing agents on the diameters of nuclear DNA-protein complexes (DPCs) from normal and neoplastic cells

Cell types compared	Molarity of NaCl used in obtaining DPCs	Diameters of normal DPCs before/after S-S reduction, μm	Diameters of neoplastic DPCs before/after S-S reduction, μm	p
NL vs. ALL	0.15	16.2/18.7	19.2/20.2	<0.005
NL vs. CLL	0.15	17.0/19.3	18.0/18.2	<0.025
NL vs. ALL	0.6	19.0/31.3	24.9/30.6	<0.005
WI38 vs. WI38-VA13	0.6	17.9/24.4	24.6/28.2	<0.05

NL = Normal lymphocytes; ALL = peripheral blood mononuclear cells from patients with acute lymphocytic leukemia; CLL = chronic lymphocytic leukemia cells. The diameters shown are the means found with 5 untreated ALL patients, 3 untreated CLL patients and 8 normal donors matched with patients for age and sex. Experiments with the normal fibroblastic cell line WI38 and the SV40 virus-transformed WI38 cells (WI38-VA13) were repeated twice. DPC 0.6s but not DPC 0.15s were treated with 300 μg/ml ethidium bromide and, in all experiments, a minimum of 200 DPCs were sized [30]. Statistics of the comparison between normal and neoplastic cells are shown.

DPCs and the possible associated cytoskeletal elements that arose from cell lyses were treated or not with S-S-reducing agents and sized by an electronic particle sizer (Coulter Channelyzer®, Fla., USA) or optically following cytocentrifugation onto glass slides [29, 30]. A summary of our experiments employing these techniques and various neoplastic human cells versus their normal counterparts is shown in table I. As seen in table I and also in figure 1, DPCs showed readily visualized decondensation following treatment with S-S-reducing agents, and to a significantly greater extent when derived from normal cells than from neoplastic cells.

To determine whether this differential decondensation of DPCs of normal versus neoplastic cells following exposure to S-S-reducing agents (fig. 1, table I) is in fact underlined by differences in the S-S bonding of their DNA-associated proteins or not, we studied the DPC 2.0s of normal and neoplastic cells by reducing and nonreducing sodium dodecyl sulfate-polyacrylamide gel electrophoresis (SDS-PAGE) [19]. Figure 2 and table II show the major protein constituents of DPC 2.0s and indicate that the protein make-up of DPC 2.0s is quite similar to what has been reported for the nuclear matrices obtained by nucleolytic digestion of nuclei [15, 16 18, 31]. It can be seen from figure 2 and table II that in normal cells the proteins undissociable from DNA by nonionic detergents plus 2.0 M NaCl are largely intermolecularly

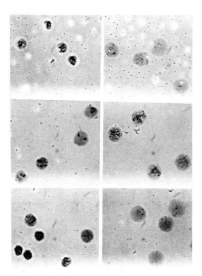

Fig. 1. Phase contrast microscopy of nuclear DNA-protein complexes obtained at 0.6 M NaCl (DPC 0.6s). Nuclear material obtained by lysis of live cells with the nonionic detergent Nonidet-40 were subjected to 0.6 M NaCl in the presence and absence of 25 mM 2-mercaptoethanol plus a positive supercoiler of DNA (ethidium bromide, 300 µg/ml), and the resulting DPC 0.6s cytocentrifuged onto glass slides [30]. Top row: DPC 0.6s from normal quiescent lymphocytes, unreduced (left) and S-S reduced (right). Middle row: DPC 0.6s from acute lymphocytic leukemia cells, unreduced (left) and S-S reduced (right). Bottom row: DPC 0.6s from normal lymphocytes cultured for 36 h with phytohemagglutinin stimulation, unreduced (left) and S-S reduced (right). Since nuclear DNA unfolds greatly after 0.6 M NaCl, visualization of DPC 0.6s by light microscopy required positive supercoiling of DNA [30]. However, the differential decondensation of normal versus leukemic DPCs by mercaptoethanol is not dependent on the ethidium bromide treatment of DPCs [30] (see also table I and fig. 4). All figures, × 400.

S-S-linked proteins. Thus the S-S-mediated condensation of DPCs appears likely to be due to intermolecular S-S bonds of more than one type of protein. Significantly, some of the S-S-linked proteins of normal DPC 2.0s could not be detected with equal amounts of DPC 2.0s from neoplastic cells, while some others of these normally intermolecularly S-S-linked proteins appeared to associate with DNA or DNA-bound molecules by nondisulfide, noncovalent bonds (fig. 2, table II). Additional alterations of the DPC 2.0 structure of neoplastic cells, discussed elsewhere [19], were present (fig. 2, table II). These SDS-PAGE findings appear related, at least mechanically, to the smaller decondensation of DPCs from neoplastic versus normal cells when exposed to S-S-reducing agents (fig. 1, table I).

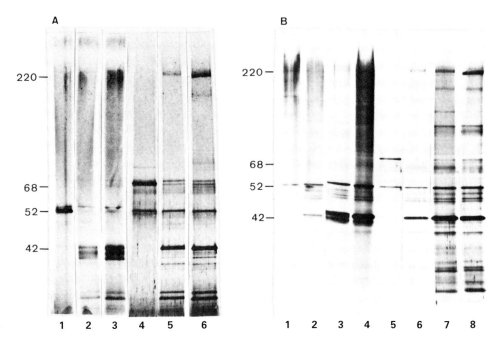

Fig. 2. SDS-PAGE of DPC 2.0s from normal and leukemic lymphocytes. Live lymphocytes were lysed in 0.8% Nonidet-40 (v/v), 0.15 M NaCl, 0.5 mM phenylmethylsulfonylfluoride, 40 mM iodoacetic acid, 18 mM EDTA, 20 mM tris, pH 7.4. After centrifugation supernatants were discarded and pelleted nuclear material resuspended in 0.15 M NaCl and then lysed in 0.5% Nonidet-40, 2.0 M NaCl, 300 µg/ml ethidium bromide, 0.5 mM phenylmethylsulfonylfluoride, 30 mM EDTA, 20 mM iodoacetic acid and 20 mM tris, pH 7.4. Lysates were loaded onto 2.7-ml cushions of 25% sucrose in SW60 Beckman polyallomer centrifuge tubes. Sucrose cushions contained 2.0 M NaCl, 20 mM iodoacetic acid and 20 mM tris, pH 7.4. Lysates were centrifuged at 44,000 rpm for 90 min at 4°C and, under such conditions, DNA-dissociated proteins remained on top of the cushion while positively supercoiled intact nuclear DNA and associated proteins pelleted. DPC 2.0s were denatured with 1.2% SDS by heating at 100°C for 10 min prior to electrophoresis. The figure was prepared by assembling selected lanes from photographs of silver-stained slab gels. Positions of molecular weight markers run on the same slabs are shown by lines molecular weight ($\times 10^{-3}$). Enzymatic digestions of DPC 2.0s prior to SDS-PAGE indicated that all of the major bands stained with silver were protein (data not shown). A1–3: DPC 2.0s derived from chronic lymphocytic leukemia cells (lane 1, 55-year-old patient), and from normal peripheral blood lymphocytes of a young adult (lane 2, 23-year-old) and old (lane 3, 61-year-old) human before S-S reduction. A4–6: Equal aliquots from solubilized DPC 2.0s of lanes A1–3 were electrophoresed after exposure to S-S-reducing agents (i.e. lane 4 shows the same material as 1; 2 as 5; and 3 as 6, except for the S-S reduction). DPC 2.0s from 4×10^6 cells were loaded to each slot. B1–4: Unreduced DPC 2.0s from 2×10^6 acute lymphocytic leukemia cells (lane 1, 14-year-old patient), from 2×10^6 normal lymphocytes of a young adult donor (lane 2, 23-year-old), from 4×10^6 normal lymphocytes of another young adult donor (lane 3, 27-year-old), and from 4×10^6 normal lymphocytes of an elderly donor (lane 4, 59-year-old). B5–8: S-S-reduced DPC 2.0s corresponding to the lanes B1–4, respectively.

Table II. Protein make-up of DPC 2.0s from normal and leukemic lymphocytes and the redox state of proteins studied with SDS-PAGE

Protein, molecular weight × 10^{-3}	Order of staining intensity of the DPC 2.0 protein from		Increase in entrance of protein into gels after S-S reduction		p
	normal lymphocytes	leukemic lymphocytes	normal lymphocytes	leukemic lymphocytes	
220	2	4	100	0	<0.001
71	6	1	100	83	
52	3	2	90	33	<0.05
42	1	3	80	50	
30	5	NP	100	NA	
15–18	4	NP	80	NA	

Order of staining intensity for each protein band was determined visually from 1 to 6 or not present (NP) with non-DNAse I-digested DPC 2.0s treated with S-S-reducing agents. For each protein band, the order of its staining intensity in 6 separate experiments (employing 12 position slab gels and 10 normal individuals plus 7 patients with acute and chronic lymphocytic leukemia) was averaged and rounded to the nearest integer. With regards to the relative staining intensities of 220,000, 71,000 and 42,000 molecular weight proteins, the leukemic cells differed significantly from normals (p <0.05 or less). Increases in the entrance of each protein into gels after S-S reduction were determined by visual comparison of protein bands from unreduced and S-S-reduced DPC 2.0s. The numbers show the percentage of cases in which an increase was seen in the staining intensity of that protein (6 separate experiments with non-DNAse I-digested DPC 2.0s). NA = Not applicable. DPC 2.0s from 2–4 × 10^6 cells/well loaded in all experiments [summarized from 19].

When SDS-PAGE profiles of DPC 2.0s from normal peripheral blood lymphocytes of young-adult humans were compared to those of elderly, greater quantities of an intermolecularly S-S-linked protein of molecular weight 220,000 appeared to associate with the DPC 2.0 DNA of older humans (fig. 2). Interestingly, earlier studies with DPC 2.0s from mouse lymphocytes had also shown greater association of an approximately 220,000 molecular weight S-S-linked protein with DNA during aging [9] (fig. 3). Precise cellular localization and function of this 220,000 molecular weight protein remains to be determined. However, the fact that the same protein band consistently failed to increase in intensity following the S-S-reducing agents treatment of DPC 2.0s from leukemic lymphocytes (fig. 2, table II) [19] stresses the importance of the S-S bonding of this protein in aging.

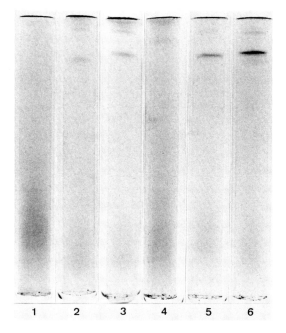

Fig. 3. SDS-PAGE of DPC 2.0s from normal splenic lymphocytes of immature, young adult and senescent mice. Nuclear material obtained by lysis of live cells with a nonionic detergent were treated with 2.0 M NaCl on top of neutral sucrose gradients in the presence of nucleolytic and proteolytic enzyme inhibitors. Intact nuclear DNA and its associated proteins were then separated from proteins dissociated by 2.0 M NaCl by centrifugation through gradients. Tubes were pierced from bottom and DPC 2.0s collected with the aid of a gradient fractionator. DPC 2.0s were then washed and pelleted, digested extensively with DNAse I, treated with 2% SDS, and loaded onto rod gels of 4.2% polyacrylamide with and without S-S reduction [9]. At all steps of DPC 2.0 preparation a free SH-blocking agent was added to media to prevent SH oxidation and SH/S-S interchange [9]. Coomassie blue staining. 1–3: Unreduced DPC 2.0s from spleen cells of immature (gel 1, 1-month-old), young adult (gel 2, 6-month-old) and senescent (gel 3, 25-month-old) mice. 4–6: Same DPC 2.0s as in gels 1–3, respectively, after S-S reduction. DPC 2.0s from 30×10^6 cells were loaded to each gel. Molecular weight of the major protein band seen was calculated to be $213,000 \mp 8,000$(SEM) from the migration of marker proteins run in parallel gels. Compared to silver staining (fig. 2), Coomassie blue staining is seen to be much less sensitive.

Ultrastructural studies of the decondensation by S-S-reducing agents of DPCs from normal and leukemic lymphocytes have also been done. In these studies we treated demembranized nuclei with 0.6 M NaCl (removes histone H1 from DNA [34]) and pelleted the resulting DPC 0.6s with and without exposure to S-S-reducing agents. Sections of pelleted DPC 0.6s were studied

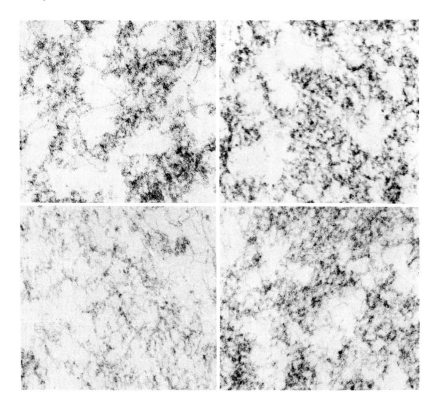

Fig. 4. Transmission electron microscopy of DPC 0.6s. DPC 0.6s obtained as described for figure 1 except with omission of ethidium bromide were pelleted, fixed with glutaraldehyde, and sections of pelleted material studied with conventional electron microscopic techniques. Top row: DPC 0.6s from normal peripheral blood lymphocytes, unreduced (left) and S-S reduced (right). Bottom row: DPC 0.6s from peripheral blood mononuclear cells of a patient with acute lymphocytic leukemia, unreduced (left) and S-S reduced (right). All figures × 80,000.

by transmission electron microscopy. Strands of DNA with associated proteins appeared to unfold to considerable lengths after treatment of normal DPC 0.6s with S-S-reducing agents (fig. 4). This unfolding was relatively less with DPC 0.6s from leukemic lymphocytes (fig. 4).

Structure of Chromatin Complex during Meiosis

At least one type, but probably most or all, of the alterations of chromatin structure associated with cellular differentiation appear reversed in meiotic cells. Thus it has been shown that the condensation and inactivation of one of the X chromatins in female embryos during early development is reversed during meiosis in female germ cells through decondensation and reactivation [35, 36]. Decondensation happens during oogonia to oocyte transition [35] and coincides with the increased DNA repair activity of meiosis [6]. Later during meiosis, histones dissociate from DNA and germ cell-specific basic proteins associate with DNA to induce transcriptional inactivity [37]. These germ cell-specific basic proteins appear to undergo extensive intermolecular S-S linking during sperm chromatin condensation in mammals [38]. The fact that the sperm chromatin decondenses after entry into the oocyte cytoplasm would suggest that at least the female germ cell has the ability to reduce the S-S bonds involved in chromatin condensation, or the involved proteins are removed from DNA by proteolytic degradation. Indeed, it now appears that oocyte is rich in free SH groups, has a reducing environment, and the chromatin decondensation therein may involve both S-S reduction and proteolytic enzymes [37]. Further, it has recently been reported that in a species phylogenetically far removed from mammals – the ciliated protozoa – the micronucleus (i.e. the germ line nucleus) shows an abundance of free SH groups but the macronucleus (somatic nucleus) of the same cell does not, except for the replicating chromatin and nucleoli [39].

Discussion

The amount of oxidative and other types of DNA damage increases during aging [40], while activities of several DNA repair processes show decreases [13, 14]. Mechanisms of these changes during aging are not known, but there is evidence that distribution of molecular lesions through DNA, as well as their repairability, may depend on the condensation state of chromatin (less condensed chromatin appears to display less damage and more efficient repair) [41, 42]. If the more condensed chromatin is less accessible to repair enzymes [43], a straightforward interpretation is that condensed chromatin structure interferes with certain types of repair, and the S-S-mediated condensation of chromatin is related to the increase of DNA damage with age. It is also of interest that the repair of oxidative damage to DNA requires reduced glutathione (GSH) [44], an endogenous S-S-reducing agent whose cellular level decreases during aging [45, 46].

The current understanding of the relationship between chromatin condensation, DNAse I accessibility of DNA in chromatin, and transcriptive activity suggests that the decreased transcriptive activity seen during aging [12] is also related to the increase of chromatin condensation with age [28].

The proliferative potential of cells also decreases during aging [2, 10, 11]. For instance following lectin stimulation of peripheral blood lymphocytes of elderly human beings, a smaller proportion of cells enters into the cell cycle and those entering spend longer times to complete the cycle in comparison to young adults [10]; furthermore antigen-stimulated lymphocytes of the elderly generate fewer progeny before ceasing proliferation [11]. Both *Makinodan and Allbright* [47] and we [9, 27] have found that the decreased proliferative potential of lymphocytes from old mice can be brought back to younger levels by addition of S-S-reducing agents to culture media. The relative enhancement of lymphocyte proliferation by thiols in these studies appeared to be greater with cells from old than young mice [9, 27, 47]. The mechanism of enhancement of cellular proliferation by S-S-reducing agents is not known at present. However, one of the targets of S-S-reducing agents in the living cell appears to be the chromatin complex [9, 27]. In fact we have now shown by SDS-PAGE of DPC 2.0s from lectin-stimulated lymphocytes that blastic transformation of quiescent lymphocytes involves partial reduction of the S-S bonds of DPC 2.0 proteins [19]. Further evidence for a partial S-S reduction in chromatin complex during blastic transformation is shown in figure 1. Diameters of unreduced DPCs from mitogen-stimulated cells show heterogeneity presumably due to the fact that in the stimulated population some cells fail to transform while others do. On the other hand, diameters of DPCs become much more uniform after S-S reduction (fig. 1). A role for the endogenous S-S-reducing agent, GSH, in blastic transformation and entrance into cell cycle of quiescent lymphocytes has been reported [48].

Although action of a single viral gene product appears to suffice to rescue cells from senescence [49], it is not known how this happens at the molecular level. *Ito* et al. [50] induced transformation of normal rat cells with the SV40 virus and gave transformed cells back to rats to determine their immunogenicity. They found that the rats developed antibodies against viral gene products as expected, and in addition against a set of cellular proteins presumed to be made antigenic by the SV40 virus transformation [50]. Note that these apparent target proteins of SV40 transformation [50] are interestingly similar to or identical with the major DPC 2.0 proteins (fig. 2, table II) in terms of their SDS-PAGE-defined sizes (except possibly for the 220,000 molecular weight species). Although this similarity might be merely a chance occurrence, evidence for the altered association of DNA with these proteins in neoplastic lymphocytes (fig. 2, table II) would suggest otherwise. Further, considering that the DNA replication and cellular differentiation involves establishment of specific associations between these proteins and DNA [20, 22, 23], an alteration of DPC 2.0 proteins-DNA association in neoplastic cells seems indeed to have the potential to modify regulations of both replication and differentiation. In fact the neoplastic cells that are induced to differentiate to normalcy have been found to display special changes in DPC 2.0 structure [23] that happen also during in vitro cellular aging [24].

How might the S-S-mediated condensation of chromatin complex increase during aging? Relevant to an answer to this question seems to be the finding that oxidation of SH groups of proteins also occurs during aging at the cell surface [51]. Decreases of GSH and the GSH/GSSG ratio during cellular aging [45, 46] might possibly be responsible for both oxidative changes, but the GSH to GSSG shift might as well be a result. If the significant GSH oxidation during differentiation of preadipocytes to adipocytes [46] can be confirmed with other cell types, the possibility should be considered that all of these SH oxidative changes are related in part to differentiation related process(es) as yet unidentified.

In summary, cell aging may occur in most part because of insufficient repair of damage to various cellular constituents by the stochastic events related to normal cellular metabolism (free radical reactions [52], sugar conjugation to proteins [53], etc.), and certain of the cellular alterations associated with cellular differentiation (e.g. chromatin condensation) may interfere with efficient repair of damage. Greater understanding of the structure-function relationship in chromatin may help to devise strategies effective in modulating cellular aging not only because chromatin complex appears to be the site of a primary aging process, but because tailored alterations of cellular metabolism (e.g. by recombining DNA) also require this knowledge.

Acknowledgements

I wish to thank Dr. *R. Hale* (Anderson Hospital, Houston, Tex.) for his help with the electron microscopy. Most of my work reviewed here was done in collaboration with and at the laboratories of Dr. *R.L. Walford* (UCLA, Los Angeles, Calif.), Dr. *B. Drewinko* (Anderson Hospital, Houston, Tex.), Dr. *L.V. Rodriguez* (Anderson Hospital, Houston, Tex.) and Dr. *J.M. Trujillo* (Anderson Hospital, Houston, Tex.).

References

1 Smith-Sonneborn, J.: DNA repair and longevity assurance in *Paramecium tetrta*. Science 203: 1115–1117 (1979).
2 Hayflick, L.: The cellular basis for biological aging; in Finch, Hayflick, Handbook of the biology of aging, pp. 159–186 (Van Nostrand Reinhold, New York 1977).
3 Röhme, D.: Evidence for a relationship between longevity of mammalian species and lifespans of normal fibroblasts in vitro and erythrocytes in vivo. Proc. natn. Acad. Sci. USA 78: 5009–5013 (1981).
4 Daniel, C.W.: Cell longevity in vivo; in Finch, Hayflick, Handbook of the biology of aging, pp. 122–158 (Van Nostrand Reinhold, New York 1977).
5 Icard, C.; Macieira-Coelho, A.: Resistance to caffeine of mouse fibroblasts after acquisition of an infinite division potential. Cell Biol. int. Rep. 5: 9–13 (1981).
6 Martin, R.: A possible genetic mechanism of aging, rejuvenation and recombination in germinal cells; in Comings, Sparkes, Fox, ICN-UCLA Symposium Proceedings, vol. 7, pp. 353–378 (Academic Press, New York 1977).
7 Taş, S.: Disulfide bonding in chromatin proteins with age and a suggested mechanism for aging and neoplasia. Exp. Gerontol. 11: 17–24 (1976).
8 Taş, S.: Involvement of disulfide bonds in the condensed structure of facultative heterochromatin and implications on cellular differentiation and aging. Gerontology 24: 358–364 (1978).
9 Taş, S.: Disulfide bonds and the structure of the chromatin complex in relation to development, aging and spontaneous malignancies of late life and modulation of structure in living cells by disulfide reducing agents (Los Angeles 1980).
10 Tice, R.R.; Schneider, E.L.; Kram, D.; Thorne, P.: Cytokinetic analysis of the impaired proliferative response of peripheral blood lymphocytes from aged humans to phytohemagglutinin. J. exp. Med. 149: 1029–1041 (1979).
11 Sohnle, P.G.; Collins-Lench, C.; Huhta, K.E.: Effect of age on generation of progeny from antigen-stimulated human lymphocytes. Mech. Age. Dev. 18: 53–58 (1982).
12 Richardson, A.; Birchenall-Sparks, M.C.; Staecker, J.L.; Hardwick, J.P.; Liu, D.S.H.: The transcription of various types of ribonucleic acid by hepatocytes isolated from rats of various ages. J. Geront. 37: 666–672 (1982).
13 Licastro, F.; Franceschi, C.; Chiricolo, M.; Battelli, M.G.; Tabacchi, P.; Cenci, M.; Barboni, F.; Pallenzona, D.: DNA repair after gamma radiation and superoxide dismutase activity in lymphocytes from subjects of far advanced age. Carcinogenesis 3: 45–48 (1982).
14 Schneider, E.L.; Gilman, B.: Sister chromatid exchanges and aging. III. The effect of donor age on mutagen induced sister chromatid exchange in human diploid fibroblasts. Hum. Genet. 46: 57–63 (1979).
15 Nakayasu, H.; Ueda, K.: Isolation and characterization of bovine lymphocyte nuclear matrix. Cell Struct. Funct. 6: 181–190 (1981).
16 Adolph, K.S.: Organization of chromosomes in HeLa cells: isolation of histone depleted nuclei and nuclear scaffolds. J. Cell Sci. 42: 291–304 (1980).
17 Cook, P.R.; Brazell, I.A.: Supercoils in human DNA. J. Cell Sci. 19: 261–279 (1975).
18 Razin, S.V.; Chernokhvostov, V.V.; Roodyn, A.V.; Zbarsky, I.B.; Georgiev, G.: Proteins tightly bound to DNA in the regions of DNA attachment to the skeletal structures of interphase nuclei and metaphase chromosomes. Cell 27: 65–73 (1981).
19 Taş, S.; Rodriguez, L.W.; Drewinko, B.; Trujillo, J.M.: Evidence for extensive intermo-

lecular disulfide bonds of the proteins of nonionic detergent-high-salt-resistant skeletons of normal lymphocytes, and the altered structure in leukemia (submitted).

20　Vogelstein, B.; Pardoll, D.M.; Coffey, D.S.: Supercoiled loops and eukaryotic DNA replication. Cell *22:* 79–85 (1980).

21　Jackson, D.A.; McCready, S.J.; Cook, P.R.: RNA is synthesized at the nuclear cage. Nature, Lond. *292:* 552–555 (1981).

22　Buongiorno-Nardelli, M.; Micheli, G.; Carri, M.T.; Marilley, M.: A relationship between replicon size and supercoiled loop domains in the eukaryotic genome. Nature, Lond. *298:* 100–102 (1982).

23　Luchnik, A.N.; Glaser, V.M.: Decrease in the number of DNA topological turns during friend leukemia differentiation. Mol. Gen. Genet. *178:* 459–463 (1980).

24　Luchnik, A.N.; Glaser, V.M.: DNA topological linking numbers in malignantly transformed syrian hamster cells. Mol. Gen. Genet. *183:* 553–556 (1981).

25　Hartwig, M.: The size of independently supercoiled domains in nuclear DNA from normal human lymphocytes and leukemic lymphoblasts. Biochim. biophys. Acta *698:* 214–217 (1982).

26　Taş, S.; Tam, C.F.; Walford, R.L.: Disulfide bonds and the structure of the chromatin complex in relation to aging. Mech. Age. Dev. *12:* 65–80 (1980).

27　Taş, S.; Walford, R.L.: Increased disulfide mediated condensation of the nuclear DNA-protein complex in lymphocytes during postnatal development and aging. Mech. Age. Dev. *19:* 73–84 (1982).

28　Taş, S.; Walford, R.L.: Influence of disulfide reducing agents on fractionation of the chromatin complex by endogenous nucleases and deoxyribonuclease I in aging mice. J. Geront. *37:* 673–679 (1982).

29　Taş, S.; Drewinko, B.; Trujillo, J.M.: Disulfide mediated condensation of the nuclear DNA-protein complex (DPC) in normal and leukemic lymphocytes. Proc. Am. Ass. Cancer Res. *23:* 11 (1982).

30　Taş, S.; Drewinko, B.; Rodriguez, L.W.; Trujillo, J.M.: Consistently greater decondensation of the nuclear DNA-protein complexes from normal lymphocytes than leukemic lymphocytes following treatment with thiol molecules (submitted).

31　Kaufmann, S.H.; Coffey, D.S.; Shaper, J.H.: Considerations in the isolation of rat liver nuclear matrix, nuclear envelope and pore complex lamina. Expl. Cell Res. *132:* 105–123 (1981).

32　Ronne, M.; Wulf, H.C.: Human cells in suspension. 3. The effect of in vitro aging on metaphase chromosome contraction in human lymphoid cells. Hum. Genet. *55:* 231–236 (1980).

33　Martinez, A.O.; Norwood, T.H.; Prothero, J.W.; Martin, G.M.: Evidence for clonal attenuation of growth potential in HeLa cells. In Vitro *14:* 996–1002 (1978).

34　Zelenin, A.V.; Vinogradova, N.G.: Influence of partial deproteinization on the cytochemical properties of DNP of lymphocyte nuclei. Expl. Cell Res. *82:* 411–414 (1973).

35　Gartler, S.M.; Rivest, M.; Cole, R.E.: Cytological evidence for an inactive X chromosome in murine oogonia. Cytogenet. Cell Genet. *28:* 203–207 (1980).

36　Kratzer, P.G.; Chapman, V.M.: X chromosome reactivation in oocytes of Mus caroli. Proc. natn. Acad. Sci. USA *78:* 3093–3097 (1981).

37　Rodman, T.C.; Pruslin, F.H.; Hoffmann, H.P.; Allfrey, V.G.: Turnover of basic chromosomal proteins in fertilized eggs: a cytoimmunochemical study of events in vivo. J. Cell Biol. *90:* 351–361 (1981).

38 Marushige, Y.; Marushige, K.: Transformation of sperm histone during formation and maturation of rat spermatozoa. J. biol. Chem. 250: 39–45 (1975).
39 Allen, R.L.; Olins, D.E.: Examination of chromatin replication in the ciliated protozoa with silver staining and thiol specific coumarin maleimide. J. Cell Biol. 95: 72a (1982).
40 Sharma, R.C.; Yamamoto, O.: Base modification in adult animal liver DNA and similarity to radiation induced base modification. Biochem. biophys. Res. Commun. 96: 662–671 (1980).
41 Sono, A.; Sakaguchi, K.: The distribution of sister chromatid exchanges and chromosomal aberrations induced by 5-fluorodeoxyuridine and ethylmethanesulfonate in the euchromatin and heterochromatin of Chinese hamster cells. Cell Struct. Funct. 5: 175–182 (1980).
42 Chiu, S.M.; Oleinick, N.L.: The sensitivity of active and inactive chromatin to ionizing radiation-induced DNA strand breakage. Int. J. Radiat. Biol. 41: 71–77 (1982).
43 Montagna, R.A.; Maizel, A.L.; Becker, F.F.; Rodriguez, L.W.: Chromatin conformation modulates repair of single strand interruptions by polynucleotide ligase ^{3}H AMP. Chem. biol. Interact. 33: 149–161 (1981).
44 Edgren, M.; Revesz, L.; Larsson, A.: Induction and repair of single strand breaks after X-irradiation of human fibroblasts deficient in glutathione. Int. J. Radiat. Biol. 40: 355–363 (1981).
45 Hazelton, G.A.; Lang, C.A.: Glutathione contents of tissues in the aging mouse. Biochem. J. 188: 25–30 (1980).
46 Takahashi, S.; Zeydel, M.: Glutamyl transpeptidase and glutathione in aging IMR-90 fibroblasts and in differentiating 3T3L1 preadipocytes. Archs Biochem. Biophys. 214: 260–267 (1982).
47 Makinodan, T.; Allbright, J.W.: Restoration of impaired immune functions in aging animals. II. Effects of mercaptoethanol in enhancing the reduced primary antibody responsiveness in vitro. Mech. Ageing Dev. 10: 325–340 (1979).
48 Fischman, C.M.; Udey, M.C.; Kurtz, M.; Wedner, H.J.: Inhibition of lectin induced lymphocyte activation by 2-cyclohexene-1-one: decreased intracellular glutathione inhibits an early event in the activation sequence. J. Immun. 127: 2257–2262 (1981).
49 Rassoulzadegan, M.; Cowie, A.; Carr, A.; Glaichenhaus, N.; Kamen, R.; Cuzin, F.: The roles of individual polyoma virus early proteins in oncogenic transformation. Nature, Lond. 300: 713–718 (1982).
50 Ito, Y.; Spurr, N.; Dulbecco, R.: Characterization of polyoma virus T antigen. Proc. natn. Acad. Sci. USA 74: 1259–1263 (1977).
51 Hughes, B.A.; Roth, G.S.; Pitha, J.: Age related decrease in repair of oxidative damage to surface sulfhydryl groups on rat adipocytes. J. Cell Physiol. 103: 349–353 (1980).
52 Harman, D.: The aging process. Proc. natn. Acad. Sci. USA 78: 7124–7128 (1981).
53 Monnier, V.M.; Cerami, A.: Nonenzymatic browning in vivo: possible process for aging of long lived proteins. Science 211: 491–493 (1981).

Dr. S. Taş, Tübitak Research Institute for Basic Sciences, PO Box 74, Gebze, Kocaeli (Turkey)

A Hypothesis for in vitro Cellular Senescence Based on the Population Dynamics of Human Diploid Fibroblasts and Somatic Cell Hybrids[1]

James R. Smith

W. Alton Jones Cell Science Center, Lake Placid, N.Y., USA

Introduction

The finite in vitro proliferative potential of human diploid fibroblasts has been widely accepted as model for the study of aging at the cellular level. A number of normal cell cultures from various species have subsequently been shown to also have a finite in vitro life span. This is in contrast to the apparently unlimited division potential of tumor-derived cell lines and cell lines that are transformed in vitro either spontaneously or by treatment with viruses or carcinogens. For the most part, the arguments for the validity of these cell culture models have been based on the observation of an inverse correlation between in vitro life span and age of donor, a correlation between in vitro life span and species life span, and decreased proliferative potential of cells derived from individuals with genetic disorders that mimic premature aging [for a review, see 6]. At the present time, the value of these cell culture models to our ultimate understanding of normal in vivo cellular aging is still uncertain. However, an understanding of the mechanisms that so stringently limit the proliferative potential of normal animal cells in culture is of significant interest in its own right. Additionally, an understanding of these processes will most likely be of relevance in the study of normal and abnormal development and the processes leading to the uncontrolled growth of tumor cells.

---

[1] This work was supported by NIH grants AG 03262, T32 AG 00043, and NIH Contract NOI AG-2-2132. I am grateful to *Cindy Gendron* for her expeditious typing.

The purpose of this paper is to describe the work of this laboratory over the past several years that has led us to our current hypothesis of the genetic control of cellular senescence.

Population Dynamics of Human Diploid Fibroblast Cultures

Hayflick and Moorhead [12] had shown that normal human cells had a limited ability to proliferate in culture, and *Hayflick* [11] proposed that they be used as a model for aging at the cellular level. A number of biochemical analyses were conducted during the first 10–15 years following the proposal. However, no biochemical defects were found that could unequivocally be shown to be the ultimate cause, or even a contributing cause, of the eventual inability of these cells to divide. Because of the apparent complexity of this problem we decided to study in detail the proliferation potential of the individual cells within mass cultures of human embryonic lung fibroblast cells (WI-38).

Hayflick [11] had reported results that suggested that all cells that could form large clones would be able to complete about the same number of population doublings (PD; ~ 50). However, it had been shown that the size of clones that individual cells could form during a 1- to 2-week growth period was very variable [22] and that an increasing proportion of cells in cultures of increasing in vitro age were not capable of division [5, 21].

In order to determine the total proliferative potential of the individual cells within a mass culture, cells were seeded onto small fragments of coverslips and those fragments to which only one cell had attached were transferred to a small culture vessel [20], allowed to grow, and then subcultured to the end of their in vitro life span. We [32] found that even in very young cultures (about 8 PD in vitro) about 50% of the cells had the ability to undergo fewer than 10 PD and that the doubling potential of the remaining 50% was very variable, ranging up to the total life span of the mass culture from which they were derived. We also found that, when the same experiment was repeated using older cultures, the percentage of cells having division potential of less than 10 PD increased with culture age and the whole distribution shifted to lower doubling potential. In these older cultures, none of the clones had a doubling potential greater than the mass culture from which they were derived. These results showed that the individual cells in the culture had some means of 'telling how old they were', and also that cellular senescence was the culmination of a large number of events, with each event being inherited by the daughter cells upon cell division. This ruled out one class of explanations for growth cessation, namely that it was due to a single event or a small number of events [13].

These data also indicated that the various biochemical studies comparing young and old cultures had to be viewed in terms of the fact that there was a high proportion of 'old' cells within 'young' cultures. This would tend to minimize or obviate the observation of any age-related changes that might exist.

Merz and Ross [21] and *Absher* et al. [1] had shown that the interdivision times of cells in fibroblast cultures were very variable. Therefore, we [10] undertook to see if the assumption of a precise counting mechanism, coupled with variable interdivision time, could account for the observed clonal variability in mass cultures. We found that, if we took into account the fact that the cells had previously undergone a certain number of PD in vivo before the culture was established, we could not rule out that a precise counting mechanism was responsible for in vitro cellular aging.

To determine if intercellular variability occurs for reasons other than variable interdivision time during in vitro cultivation, we undertook to determine the variation in life span of cells which were descendants of a single cell (i.e., intraclonal variation). Since there had been a widespread belief that human diploid fibroblasts do not grow as well in clonal culture as in mass culture, we first had to demonstrate that data derived from clonal analysis could be used, with confidence, to infer the behavior of cells in mass cultures. We, therefore, experimentally showed that the PD time, total in vitro life span, and percentage of nondividing cells were essentially the same for cells in clonal culture and in mass culture [3].

To study the intraclonal variation in proliferative potential, we chose a long-lived clone of human diploid fibroblasts and isolated about 200 single cells at different points in its in vitro life span (16, 26, 36 PD after initiation of the clone) [34]. We found that within 16 PD of initiation of the clone tremendous variability had developed in the growth potential of the cells within the clone, with cells ranging from 0 to 33 PD in growth potential. The doubling potentials were distributed in a distinctly bimodal fashion, consisting of a low doubling potential mode comprised of less than 10% of the cells and a higher doubling potential mode comprised of cells ranging from 11 to 33 PD. When the clone was sampled at points nearer the end of its in vitro life span, the proportion of cells in the lower mode increased, with a concomitant decrease in the proportion and the doubling potential of cells in the higher mode. We also looked at the difference in doubling potential of cells derived from single mitotic events and found that they could differ by at least 8 PD in their proliferative capacity. These results indicate that at cell division the factors determining the longevity of cells are not necessarily divided equally between daughter cells. This unequal distribution is responsible for a major part of the heterogeneity of clonal life span observed in mass and clonal culture, and variable interdivision time contributes minimally to the heterogeneity.

From the data on subclonal variation it seems clear that at least two qualitatively different kinds of processes are occurring. One series of events gradually decreases the proliferative potential of a cell. In addition, it is possible for a single event to occur which can dramatically decrease the life span of a cell (i.e., transfer it from a high to a low doubling potential). These data are detailed enough to allow one to test various hypotheses and models that have been offered to explain cell division cessation. It is clear that a precise counting mechanism alone cannot account for the observed intraclonal variability. Several other theories and models are also not compatible with these data. Among these are the commitment theory [15, 18], the mortalization theory [30], and the model of *Prothero and Gallant* [29]. We have developed a working hypothesis based on these results and other results obtained from our laboratory and from other laboratories. The hypothesis – and its test by computer simulation [17] – will be discussed at the end of this paper.

Cell Hybridization Studies

The first indication that a specific substance could be responsible for the division cessation of normal cells came from the cell fusion studies of *Norwood* et al. [25], *Stein and Yanishevsky* [35], *Stein* et al. [36], and *Yanishevsky and Stein* [37]. In studies of DNA synthesis occurring in the nuclei of heterodikaryons within 72 h postfusion, they found the following: (1) Senescent cells could inhibit DNA synthesis in the nuclei of young cells unless the young cell was in S phase at the time of fusion; (2) senescent cell nuclei could be induced to synthesize DNA by fusion with SV40-transformed cells or HeLa cells; (3) senescent cells could inhibit DNA synthesis in the nuclei of some immortal cell lines (e.g., T98G, HT1080, SUSMI). These results led to the development of the hypothesis that senescent cells produce an inhibitor of initiation of DNA synthesis that is transmitted through the cytoplasm and can affect the nuclei of young normal and some immortal cells. SV40-transformed and HeLa cells are either unaffected by this inhibitor or can negate it, and simultaneously produce initiators of DNA synthesis that cause senescent cell nuclei to synthesize DNA. This inhibitor of DNA synthesis initiation could be involved in the eventual division cessation of cells.

Experiments involving the fusion of enucleated cytoplasts derived from senescent cells with young cells provide additional evidence for the idea that division cessation may result from the production of a substance that blocks the initiation of DNA synthesis. Senescent cytoplasts block initiation of DNA synthesis in about 50% of the fusion products [4, 7]. We have found that treatment of the senescent cytoplasts with an inhibitor of protein synthesis (cycloheximide) for 2 h or more before fusion almost completely abolishes the ability of the senescent cytoplasts to block DNA synthesis [*Drescher-Lincoln and Smith*, in preparation]. This indicates that the senescent cells are producing a protein that actively participates in blocking the initiation of DNA synthesis. If the senescent cytoplasts are treated for a short time with cycloheximide and the cytoplasts allowed to synthesize protein for at least 2 h, by removal of the cycloheximide before fusion, they regain the ability to block DNA synthesis. These results are consistent with those of *Burmer* et al. [3] who found that brief treatment of young × senescent cell fusion products at the time of fusion with protein synthesis inhibitors temporarily abolished the ability of senescent cells to block initiation of DNA synthesis.

The idea that senescent cells produce a protein that actively blocks initiation of DNA synthesis prompted us to compare the two-dimensional gel protein profiles of young and old human diploid fibroblasts. Because we are well aware of the heterogeneity of life span of mass cultures, we used clones in these studies to maximize the probability of identifying proteins that might be unique to senescent cells. To make sure that young and senescent cells were in the same stage of the cell cycle, we looked at the protein profiles of cells that had been maintained in medium with 0.3% serum for 72 h. This condition arrests young cells in the G_1 phase of the cell cycle, which is the same phase in which the majority of senescent cells are blocked. Low serum maintenance also minimized protein differences that might occur due to different serum concentrations.

To date we have identified two polypeptides in the two-dimensional gel profiles of ^{35}S-methionine-labelled proteins of senescent cells that do not appear in the polypeptide profiles of young cells [*Lincoln and Smith*, submitted]. Both have about the same molecular weight as actin with higher pIs. The short half-life of one of these proteins, as determined by pulse-chase experiments, is about the same as the effective half-life of the inhibitor of DNA synthesis present in senescent cytoplasts, discussed earlier. We are sufficiently encouraged by these results to isolate and characterize these polypeptides to determine if either of them is indeed the inhibitor of DNA synthesis found in senescent cells.

In addition to short-term studies on the control of DNA synthesis by cytoplasmic factors, investigators have also studied the contribution of nuclear and cytoplasmic components to the control of long-term proliferative potential. *Muggleton-Harris and Hayflick* [23] by micromanipulation demonstrated the importance of both the nucleus and cytoplasm in the regulation of cell proliferation. *Muggleton-Harris and Desimone* [24] studied the proliferative behavior of hybrids resulting from normal × SV40-transformed cells by micromanipulating cells and they found that the majority of the hybrids had limited proliferative potential. However, the technique restricted the number of clones that could be analyzed, and continued subculture of extensively proliferating hybrids was not done. It was assumed that such hybrids would divide indefinitely. *Bunn and Tarrant* [2], using a biochemical selection system, analyzed the proliferative potential of extensively proliferating hybrids resulting from the fusion of normal human fibroblasts and HeLa cells. They found that the hybrid clones ceased division even after prolonged proliferation and that rare immortal variant cells arose at low frequencies in the population of hybrid cells. These two studies indicated that the phenotype of finite proliferative potential was dominant in normal-immortal cell crosses. *Hoehn* et al. [14] studied hybrids resulting from the fusion of mass cultures having long or short total in vitro life spans, in order to determine the behavior of the phenotype of limited division in hybrids. Their hybrid analysis was restricted to extensively proliferating clones, as hybrids were identified by flow microfluorimetry and isoenzyme analysis of glucose-6-phosphate dehydrogenase. They concluded that the life spans of hybrids formed by the fusion of cells with low and high doubling potential were intermediate between those of hybrids formed from the fusion of cells with high doubling potential and those from the fusion of cells with low doubling potential. However, to draw conclusions on dominance and recessiveness of a phenotype we felt that not only should large numbers of hybrids be analyzed, but that analysis of hybrid life spans had to include study of clones having both low and extensive proliferative potential. The culture conditions we had already optimized for growth of normal cells at very low densities allowed us to obtain the latter information for such an analysis, if we could use a biochemical selection system for hybridization isolation.

We were fortunate in isolating an exceptionally long-lived ($\sim$ 100 PD) cell line of normal human fetal lung fibroblasts [8]. We used this cell line to obtain a clone resistant to ouabain (O^R) and thioguanine (TG^R). The mutant clone that was $O^R TG^R$ had 25 PD remaining (PDR) at the time it could be used in fusions. Such a clone could then be fused with a wild type cell and the hybrids selected for in medium containing HAT (hypoxanthine, aminopterin, thymidine) [19] and ouabain.

When we began our hybridization experiments, we used this $O^R TG^R$ clone as one parent and fused it to other clones of normal, SV40-transformed and tumor-derived cells. We used clones as the parents of the fusions, since the life span distributions of cells within clones are more homogenous than those of mass cultures. This allowed a more detailed comparison of hybrid life span distributions with those of the parents than would have been possible if mass cultures were used.

Normal × Normal Cell Fusions

With our complete analysis of hybrids unable to divide, those that could divide a few times, and those that could divide extensively, we found that the distribution of proliferative potentials of hybrids resembled that of the parent having smaller proliferative potential [27]. This indicated that the phenotype of low proliferative potential was dominant in hybrids.

From our results, it appeared that the dominance of lower proliferative potential was most closely associated with the group of cells, present in the parent population, that could achieve less than 8 PD. Further we found that when senescent clones (90–95% nondividing cells) were fused with very young cells (≥ 80 PDR), 50% of the hybrids could divide at least once. Thus it appears that in synkaryons, i.e., hybrids in which the nuclei are fused, the dominance of the senescent nucleus can be temporarily reversed by the very young nucleus.

The results obtained from fusion of very young cells with old or senescent cells were particularly interesting. They indicated the possibility that a 'dosage' phenomenon is involved in the reversal of the senescent (nondividing) phenotype. Thus, whenever very young cells (80 PDR) were fused to late phase III cells, they could reverse senescence in 50% of the nondividing cells and give rise to hybrids that would divide at least once. However, when the young cells had only 40 PDR, they were unable to reverse senescence in late phase III cultures. The results suggested that young cells contain inducers of DNA synthesis that decrease as the culture completes its proliferative potential and that a certain threshold of inducers is required to reinitiate DNA synthesis in late phase III cells. The results also suggested that the threshold for reinitiation of DNA synthesis is different in nondividing cells present in old (PDR 1), early and late phase III cultures. Thus, young cells with 40 PDR could not reinitiate DNA synthesis in late phase III cells after fusion, but could do so in old (PDR 1) cultures. The reason for this difference between nondividing cells in old versus senescent cultures could be that, with time postsenescence, cells are altered in some way, possibly at the site for initiation of DNA synthesis, and cannot be as easily induced to divide.

We have found that, following fusion with SV40-transformed cells, DNA synthesis is initiated in 100% of the cells present in a culture of normal human fibroblasts that had been maintained for 6 months postsenescence [26]. If inducers of DNA synthesis are indeed involved in the process of DNA synthesis induction, this suggests that transformed cells have a very high level of, or more efficient, inducers than young cells. It is also possible that the integrated viral genome acts in some way to override the blockage of DNA synthesis initiation in senescent cells.

Normal × Immortal Cell Fusions

We began these studies with crosses between normal human fibroblasts and SV40-transformed cells [26]. We obtained results similar to those of *Muggleton-Harris and Desimone* [24] in that the majority of the hybrids (70%) had very limited proliferative potential (<8 PD). The remaining 30% of the clones had a variable range of doubling potential (16–69 PD), but all clones eventually ceased division. They [24] had assumed that such clones would divide indefinitely. Following division cessation, foci of dividing cells occurred in about half the hybrid clones. These divided to repopulate the culture vessel, could be subcultured, and then would divide indefinitely. The foci appeared at very low frequencies of ~ 1 in 10^5 cells. *Bunn and Tarrant* [2] had observed a similar pattern of in vitro behavior in extensively proliferating hybrids resulting from fusion of normal cells with HeLa cells.

The possibility existed that the hybrids from these normal-immortal cell crosses ceased division because of loss of some part of the immortal genome. We had used SV40-transformed cells because a gene product of the viral A gene region can be easily assayed. We tested our normal × SV40-transformed hybrid clones for viral T antigen by indirect immunofluorescence. Clones capable of <6 PD as well as those capable of extensive proliferation were assayed and found to be T antigen positive. Further, we did Southern blot analysis of EcoRI-restricted DNA isolated from hybrids before division cessation, and after division had resumed, following formation of foci of dividing cells. The DNA band patterns from such hybrids and the SV40-transformed parent used for fusion were essentially the same. This clearly demonstrated that the phenotype of limited division was dominant over the immortal phenotype in the presence of a stably integrated viral genome that was being expressed. Thus the phenotype of immortality was recessive in fusions involving normal cells with SV40-transformed or HeLa cells.

We proceeded to examine the generality of this phenomenon by fusing normal cells with other immortal human cell lines: HT1080 (derived from a fibrosarcoma), 108021A (an APRT$^-$ derivative of HT1080), 143BTK$^-$ (an RNA virus-infected cell line derived from a osteosarcoma), T98G (an isolate from a glioblastoma) [28]. We found that in all cases the majority of the hybrids (80–90%) had very limited proliferative potential. Again all extensively proliferating clones eventually ceased division even after achieving as many as 68 PD. Thus the phenotype of immortality was recessive in hybrids involving the fusion of normal human cells with quite a range of immortal human cell lines.

The data indicated that immortality results from changes or dysfunctions in the normal cell genome that are recessive in hybrids. These lost functions are apparently contributed by a normal cell genome, causing hybrids from such fusions to be similar to normal cells in their proliferative behavior. The next question was whether immortality could result from more than one such change or dysfunction. If this were true, fusion of immortal cell lines with each other might lead to complementation of these changes to yield hybrids having limited division potential.

Fusion of SV40-Transformed Cells with Other Immortal Cells

We fused the same group of immortal cell lines we had previously used (HeLa, HT1080, 108021A, 143BTK$^-$, T98G) with an SV40-transformed double mutant cell line we had derived. We also fused this parent SV40-transformed line with the wild type clone from which it was derived and two other independently derived SV40-transformed cell lines [28]. In the fusions among the different SV40-transformed lines, there was no complementation to yield hybrids with limited life span. All hybrids could divide indefinitely. This indicates that, in the three SV40-transformed cell lines we studied, immortality is the result of either the same or noncomplementary genetic defects. In all the other fusions, the majority of the hybrid clones were capable of very limited division (< 8 PD).

Since highly aneuploid parents were used in the fusions, the possibility existed that some events other than true complementation resulted in hybrids with low proliferative potentials. We, therefore, based our major conclusions on the results obtained with extensively proliferating hybrid clones. When

HT1080 or its subclone were fused with SV40-transformed human fibroblasts, the hybrids that achieved more than 8 PD could divide indefinitely and exhibited no complementation for mortality. In all other cases there was complementation and all hybrid clones analyzed did cease division. In the case of fusions of HeLa and T98G cells with SV40-transformed cells, there was resumption of division in some instances following the appearance of foci of dividing cells in the hybrids. Fusion with 143BTK$^-$ cells yielded no clones capable of indefinite proliferation. All clones that had limited division potential stained positive for SV40 T antigen by indirect immunofluorescence.

Our conclusion from this work is that the phenotype of immortality is recessive in hybrids with normal human cells and that complementation between immortal parent cells can occur to result in hybrids with limited division potential. The data indicate that two or more different events (or sets of events) can result in immortality and that these events affect the normal cell growth control mechanisms.

A Possible Mechanism of Cellular Senescence

If one accepts the idea that, at the end of their in vitro life span, cells produce a substance that can diffuse through the cytoplasm and prevent the initiation of DNA synthesis, then one must ask how the timing of the production of this inhibitor is controlled. The facts that we must take into account before proposing a mechanism for in vitro cellular aging are listed below.

(1) Some cells in cultures of human diploid fibroblasts probably go through as many as 200 divisions before division ceases [9].
(2) Cells somehow count the number of previous divisions they have achieved.
(3) At division, the doubling potentials of the daughter cells often differ by several population doublings.
(4) The doubling potential of cells decreases in two qualitatively different ways: (i) a gradual decrease at each division and (ii) an abrupt transition of cells from a relatively large doubling potential to cells with a small doubling potential (≤ 8 PD). The probability of this transition increases as cells approach the end of their in vitro life span [34].
(5) Cells can become immortal at very low frequencies either by viral or carcinogen treatment, or spontaneously in vivo by tumor formation.
(6) Heterodikaryon experiments suggest that cells can become immortal by

at least two different mechanisms [25, 36] and our cell hybridization experiments indicate two different complementation groups for cellular immortality.

(7) In hybrid cells formed by the fusion of immortal cells and normal cells, the phenotype of finite proliferative potential is dominant.

We have developed a working hypothesis for in vitro cellular senescence that takes into consideration all of the points listed above [33]. The hypothesis has been reduced to a mathematical formulation [16] and tested by computer simulation [16] against our intraclonal life span distribution data [34] and against several other parameters of population dynamics of cellular growth and senescence. The key elements of the hypothesis are:

(1) Normal diploid cells contain the genetic information to produce a substance that blocks the initiation of DNA synthesis (inhibitor).

(2) This genetic information is not expressed until the cell has completed its in vitro life span, that is the inhibitor gene(s) is repressed in growing cells.

(3) The efficiency with which the inhibitor gene is repressed decreases with increasing PD level.

In order to make our hypothesis conceptually simple and easy to model, we assumed that the genes coding for the repressor are present in multiple copies. The loss of one or more copies of the repressor gene provides a heritable timing or counting mechanism for cells to 'tell how old they are' and unequal distribution of the repressor gene copies at division provides a mechanism for the observed unequal distribution of life span at cell division.

To provide an aging component to this model, it is necessary that the average number of active repressor genes per cell decreases at each cell division. We have assumed that this inactivation occurs by a process that requires the proper functioning of the normal cell genetic program. For the purposes of discussion, we have called this process 'suppression'. Finally, in order to account for the bimodal distribution of subclonal life span we assumed that the probability of transferring to the low doubling potential domain was a function of the number of active repressor gene copies present. We proposed that the smaller the number of active gene copies, the lower the concentration of repressor in the cell and the more likely that the inhibitor gene would be expressed. The fact that in heterokaryon experiments the repressors from the young cell cannot shut off production of the senescent cell inhibition indicates that the inhibitor gene product is 'locked on' by a positive feedback process. In our model we have assumed that the repressor and inhibitor compete for the same binding site or that the inhibitor interacts with the repressor to inactivate it. Therefore, if repression fails momentarily and some inhibitor is produced, the cells enter the low doubling potential mode where the inhibitor concentration increases to the point that the cell can no longer divide.

Using this model we were able to obtain, by computer simulation, a very good fit to the data on intraclonal life span distributions [34] and other growth parameters of human diploid fibroblasts. In obtaining the fit to the intraclonal data, we discovered several interesting properties of the model. We found that the process of 'suppression' had to inactivate a fixed *number* of repressor gene copies and not a fixed *proportion* of gene copies at each cell generation. This is of interest because it implies that the loss of repressor gene copies occurs as a genetically programmed process as distinct from a stochastic process. Furthermore, a disruption in the process of 'suppression' would result in immortal cell lines. These immortal cell lines could have a large number of repressor gene copies and could possibly induce DNA synthesis in heterokaryons with senescent cells, such as that observed with SV40-transformed cells or HeLa cells. However, in hybrids between normal cells (which have an active 'suppression' mechanism) and this class of immortal cells the hybrids should have a finite life span. This is what we have observed in our cell hybrid experiments [26].

Another way in which cells could become immortal would be to lose the ability to produce the inhibitor (e.g., by mutation of the inhibitor gene). In this case one would expect that upon fusion with senescent cells initiation of DNA synthesis would be blocked in the nucleus of the immortal cell. This is the observed result in heterodikaryon experiments involving the immortal cell lines T98G, HT1080, and SUSMI with senescent cells [36]. If the disruption of two different processes involved in determining the limited in vitro life span of normal cells can each lead to immortal cell lines, then we would expect that hybrid cells created by the fusion of two such different immortal cells might have a limited in vitro life span. As discussed above, we have obtained this result from hybrids between several immortal cell lines [28].

One other feature of the model should be mentioned: the distribution of repressor gene copies between daughter cells at division. We found that a completely random distribution (such as would result from the distribution of completely independent particles) resulted in intraclonal distributions much broader than those observed. In order to fit the observed data, the number of repressor gene copies received by sister cells differed by 5 or 6 at the most, with the most probable event being equal distribution. Thus the repressor gene copies are most likely chromosomal, and processes such as sister chromatid exchange could possibly account for the extent of unequal distribution observed.

While this hypothesis does not provide detailed biochemical mechanisms for the processes involved, it does give us a good idea of the kinds of processes that must be involved in determining the finite life span of normal cells. An attractive feature of this hypothesis is its compatibility with a wide range of experimental data from aging and immortal cell populations.

References

1 Absher, P.M.; Absher, R.G.; Barnes, W.D.: Genealogies of clones of diploid fibroblasts. Cinemicrophotographic observations of cell division patterns in relation to population age. Expl Cell Res. *88:* 95–104 (1974).
2 Bunn, C.L.; Tarrant, G.M.: Limited life span in somatic cell hybrids and cybrids. Expl Cell Res. *127:* 385–396 (1980).
3 Burmer, G.C.; Zeigler, C.J.; Norwood, T.H.: Evidence for endogenous polypeptide-mediated inhibition of cell cycle transit in human diploid cells. J. Cell Biol. *94:* 187–192 (1982).
4 Burmer, G.C.; Motulsky, H.; Zeigler, C.J.; Norwood, T.H.: Inhibition of DNA synthesis of young cycling human diploid fibroblast-like cells upon fusion to enucleate cytoplasts from senescent cells. Expl Cell Res. *145:* 79–84 (1983).
5 Cristofalo, V.J.; Sharf, B.B.: Cellular senescence and DNA synthesis. Expl Cell Res. *76:* 419–427 (1973).
6 Cristofalo, V.J.; Stanulis-Praeger, B.M.: Cellular senescence in vitro; in Maramorosch, Advances in cell culture, vol. 2, pp. 1–68 (Academic Press, New York 1982).
7 Drescher-Lincoln, C.K.; Smith, J.R.: Inhibition of DNA synthesis in proliferating human diploid fibroblasts by fusion with senescent cytoplasts. Expl Cell Res. *144:* 455–462 (1983).
8 Duthu, G.S.; Braunschweiger, K.I.; Pereira-Smith, O.M.; Norwood, T.H.; Smith, J.R.: A long-lived human diploid fibroblast line for cellular aging studies: applications in cell hybridization. Mech. Age. Dev. *20:* 243–252 (1982).
9 Good, P.I.: Subcultivations, splits, doublings, and generations in cultures of human diploid fibroblasts. Cell Tiss. Kinet. *5:* 319–323 (1972).
10 Good, P.I.; Smith, J.R.: Age distribution of human diploid fibroblasts. A stochastic model for in vitro aging. Biophys. J. *14:* 811–823 (1974).
11 Hayflick, L.: The limited in vitro lifetime of human diploid cell strains. Expl Cell Res. *37:* 614–636 (1965).
12 Hayflick, L.; Moorhead, P.S.: The serial cultivation of human diploid cell strains. Expl Cell Res. *25:* 585–621 (1961).
13 Hirsch, H.R.; Curtis, H.J.: Dynamics of growth in mammalian diploid tissue cultures. J. theor. Biol. *42:* 227–244 (1973).
14 Hoehn, H.; Bryant, E.M.; Martin, G.M.: The replicative life spans of euploid hybrids derived from short-lived and long-lived human skin fibroblast cultures. Cytogenet. Cell Genet. *21:* 349–362 (1974).
15 Holliday, R.; Huschtscha, L.I.; Tarrant, G.M.; Kirkwood, T.B.L.: Testing the commitment theory of cellular aging. The finite lifespans of human fibroblasts may be due to the decline and loss of a subpopulation of immortal cells. Science *198:* 366–372 (1977).

16 Jones, R.B.; Lumpkin, C.K.; Smith, J.R.: A stochastic model for cellular senescence. Part I. Theoretical considerations. J. theor. Biol. *86:* 581–592 (1980).

17 Jones, R.G.; Smith, J.R.: A stochastic model of cellular senescence. II. Concordance with experimental data. J. theor. Biol. *96:* 443–460 (1982).

18 Kirkwood, T.B.L.; Holliday, R.: Commitment to senescence: a model for the finite and infinite growth of diploid and transformed human fibroblasts in culture. J. theor. Biol. *53:* 481–496 (1975).

19 Littlefield, J.W.: Selection of hybrids from matings of fibroblasts in vitro and their presumed recombinants. Science *145:* 409 (1964).

20 Martin, G.M.; Taun, A.: A definitive cloning technique for human fibroblast cultures. Proc. Soc. exp. Biol. Med. *123:* 138–140 (1966).

21 Merz, G.S.; Ross, J.D.: Viability of human diploid cells as a function of in vitro age. J. cell. Physiol. *74:* 219–222 (1969).

22 Merz, G.S., Jr.; Ross, J.D.: Clone size variation in the human diploid cell strain, WI-38. J. cell. Physiol. *82:* 75–80 (1973).

23 Muggleton-Harris, A.L.; Hayflick, L.: Cellular aging studied by the reconstruction of replicating cells from nuclei and cytoplasms isolated from normal human diploid cells. Expl Cell Res. *103:* 321–330 (1976).

24 Muggleton-Harris, A.L.; Desimone, D.W.: Replicative potentials of various fusion products between WI-38 and SV40 transformed WI-38 cells and their components. Somatic Cell Genet. *6:* 689–698 (1980).

25 Norwood, T.H.; Pendergrass, W.R.; Sprague, C.A.; Martin, G.M.: Dominance of the senescent phenotype in heterokaryons between replicative and postreplicative human fibroblast-like cells. Proc. natn. Acad. Sci. USA *71:* 2231–2235 (1974).

26 Pereira-Smith, O.M.; Smith, J.R.: Expression of SV40 T antigen in finite life span hybrids of normal and SV40 transformed fibroblasts. Somatic Cell Genet. *7:* 411–421 (1981).

27 Pereira-Smith, O.M.; Smith, J.R.: Phenotype of low proliferative potential is dominant in hybrids of normal human fibroblasts. Somatic Cell Genet. *8:* 731–742 (1982).

28 Pereira-Smith, O.M.; Smith, J.R.: Evidence for the recessive nature of cellular immortality. Science *221:* 964–966 (1983).

29 Prothero, J.; Gallant, J.A.: A model of clonal attenuation. Proc. natn. Acad. Sci. USA *78:* 333–337 (1981).

30 Shall, S.; Stein, W.D.: A mortalization theory for the control of cell proliferation and for the origin of immortal cell lines. J. theor. Biol. *76:* 219–231 (1979).

31 Smith, J.R.; Braunschweiger, K.I.: Growth of human embryonic fibroblasts at clonal density: concordance with results from mass cultures. J. cell. Physiol. *98:* 597–602 (1979).

32 Smith, J.R.; Hayflick, L.: Variation in the life span of clones derived from human diploid cell strains. J. Cell Biol. *62:* 48–53 (1974).

33 Smith, J.R.; Lumpkin, C.K., Jr.: Loss of gene repression activity: a theory of cellular senescence. Mech. Age. Dev. *13:* 387–392 (1980).

34 Smith, J.R.; Whitney, R.G.: Intraclonal variation in proliferative potential of human diploid fibroblasts: stochastic mechanisms for cellular aging. Science *207:* 82–84 (1980).

35 Stein, G.H.; Yanishevsky, R.M.: Entry into S phase is inhibited in two immortal cell lines fused to senescent human diploid cells. Expl Cell Res. *120:* 155–165 (1979).

36 Stein, G.H.; Yanishevsky, R.M.; Gordon, L.; Beeson, M.: Carcinogen transformed human cells are inhibited from entry into S phase by fusion to senescent cells, but cells transformed

by DNA tumor viruses overcome the inhibition. Proc. natn. Acad. Sci. USA *79:* 5287–5291 (1982).

37 Yanishevsky, R.M.; Stein, G.H.: Ongoing DNA synthesis continues in young human diploid cells (HDC) fused to senescent HDC, but entry into S phase is inhibited. Expl Cell Res. *126:* 469–472 (1980).

J.R. Smith, PhD, Department of Virology and Epidemiology, Baylor College of Medicine, Houston, TX 77030 (USA)

V. Cytoplasmic Ageing

Mitochondrial Alterations in Ageing Mouse Neuroblastoma Cells in Culture

P.E. Spoerri[1]

Department of Anatomy, Georg-August-Universität, Göttingen, FRG

Introduction

One of the most conspicuous changes in the tissues of the ageing animals is the presence of the intracellular ageing pigment, lipofuscin, which can be seen by both light and light and electron microscopy in tissues. The age-related accumulation of this material has been noted in a variety of cells [*Strehler* et al., 1959; *Strehler*, 1964; *Samorajsky* et al., 1965; *Bourne*, 1973; *Brizzee* et al., 1975].

Despite a wealth of data most ultrastructural studies on ageing of postmitotic material have concentrated on the accumulation of dense bodies including lipofuscin and have discussed the possible origins of such structures [*Hasan and Glees*, 1972; *Travis and Travis*, 1972; *Tomanek and Karlsson*, 1973; *Hasan* et al., 1974; *Glees and Hasan*, 1976; *Koobs* et al., 1978]. However, there is little comment in the literature on the possible changes in the structure of mitochondria, with the exception of the histochemical biochemical and morphological report of *Colcolough* et al. [1972], and the in vitro and in vivo studies carried out in our laboratory [*Glees and Gopinath*, 1973; *Spoerri and Glees*, 1973, 1974, 1979; *Glees* et al., 1974, 1975; *Spoerri* et al., 1974; *Ahmed and Glees*, 1976; *Spoerri*, 1982].

Intrinsic mitochondrial senescence may be related to age-related changes in other organelles and thus be detrimental to the cell [*Miquel* et al., 1980]. On the other hand mitochondria may be directly involved in the gen-

---

[1] I am grateful for the technical assistance of Ms. *Elke Heyder* and the photographic work of Mr. *Rod Dungan*. This work was supported by the German Science Foundation (DFG grant Sp 237/2–1).

esis of lipofuscin as we have previously reported and discussed. The present study once more emphasizes mitochondrial alterations as seen in aged mouse neuroblastoma cultures.

Materials and Methods

The clonal line Neuro-2a, C1300 mouse neuroblastoma, was purchased from the American Type Collection Culture, Rockville, Md. Cultures were grown at 37 °C in Eagle's minimum essential medium supplemented with 10% calf serum. In Falcon flasks, the monolayer cultures were fed twice to three times a week. After 5, 10, 15, 30, 38, 45, 52, 60 and 68 days in vitro the cells were fixed for electron microscopy. After draining off the medium the cells were fixed in situ at 36 °C for 15 min using 3% glutaraldehyde in phosphate buffer (pH 6.8). Then the cells were postfixed in 2% osmium tetroxide in 0.2 M phosphate-buffered saccharose for 30 min. After dehydration in ethanol the cultures were embedded in Epon. Selected cells were mounted and cut as described in detail elsewhere [*Spoerri* et al., 1980a]. Ultrathin sections for electron microscopy were stained on grids with a saturated aqueous solution of uranyl acetate and lead citrate. For light microscopy, 0.5- to 1.0 μm thick sections were stained with toluidine blue.

Results

Osmiophilic bodies identified as lipofuscin are present in all of the neuron-like cells of different ages examined [*Spoerri* et al., 1980b], but their number varies depending on the age of the culture, as assessed by the comparative study of a large number of sections and electron micrographs. The mitochondria of the senescent neurons from 30-day-old cultures and onwards (fig. 1–12) possess a swollen granular matrix and show a loss of cristae, vacuolization of the matrix and disruption of their membranes. Or appear condensed with a high electron density. The altered mitochondria are found mostly adjacent or in the vicinity of the dense lipofuscin granules.

The most striking feature of the perikarya of the aged neurons is the progressive accumulation of osmiophilic bodies. Their number but not their size is considerably increased when compared to those present in the younger cells (fig. 1–9). In particular considerable variations in the morphology and electron density of the pigment bodies in the different neurons exist. Some are vacuolated showing electron lucid areas (fig. 1–9), and others are non-vacuolated (fig. 6, 10–12). Other dense bodies are clear at one point and show concentric dense structures somewhere else (fig. 1, 7–9). Some lipofuscin granules basically resemble mitochondria by showing double delimiting membranes (fig. 1–12), while others depict remnants of cristae within the

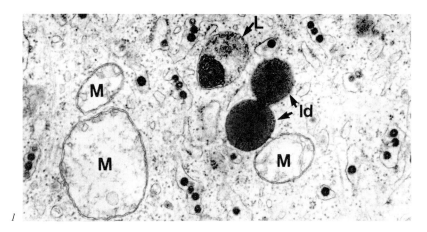
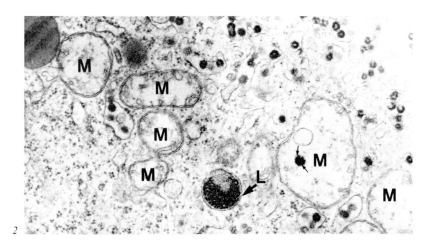
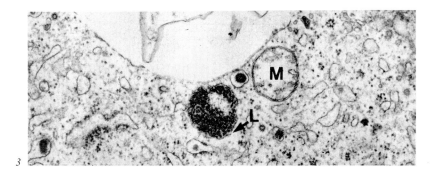

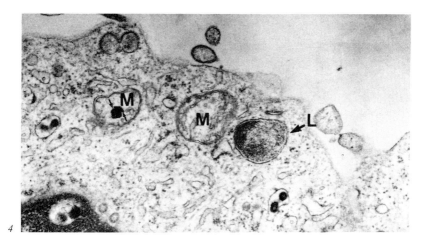

Fig. 1–4. Profiles of neurons from 30-day-old mouse neuroblastoma cultures depicting degenerated mitochondria (M) and lipofuscin granules (L). Note the swollen mitochondria (M) with granular matrix and loss of cristae. In figures 2 and 4, the intramitochondrial osmiophilic bodies (arrows) are prominent. The lipofuscin granules (L) show double membranes and peripherally located pigment. Note the concentric cristae-like structures in figure 1 and the remnants of cristae within the electron-lucent part of the body in figure 4. In figure 1, ld = lipid droplets. × 20,000.

dark matrix (fig. 3–5, 7, 9). No special relationship of the dense bodies to other organelles such as endoplasmic reticulum, Golgi, and lysosomes can be observed.

Discussion

The data suggest that ageing in mouse neuroblastoma cells in culture is associated with fine structural alterations in the mitochondria. As a consequence there can only be a decrease in the number of these organelles. It appears well established that there is an age-related decrease in the number of mitochondria present in fixed postmitotic cells and other cells [*Tauchi and Sato*, 1968; *Miquel* et al., 1980]. During the course of physiological respiration which occurs in the inner mitochondrial membranes or cristae, free radicals are formed during the univalent reduction of oxygen in the respiratory chain. Profound and destructive molecular changes occur from successive oxidation of structures rich in unsaturated lipids. The most vulnerable are mitochondrial and microsomal membranes. It has been suggested that free

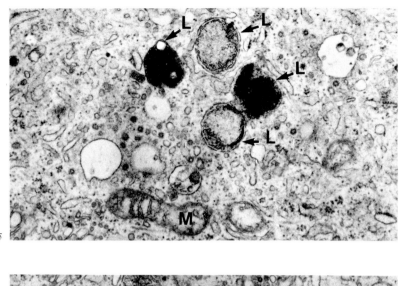

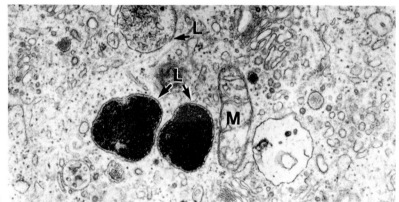

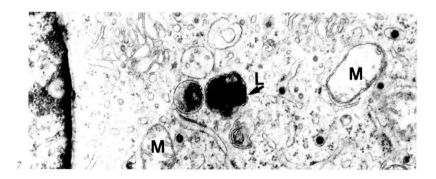

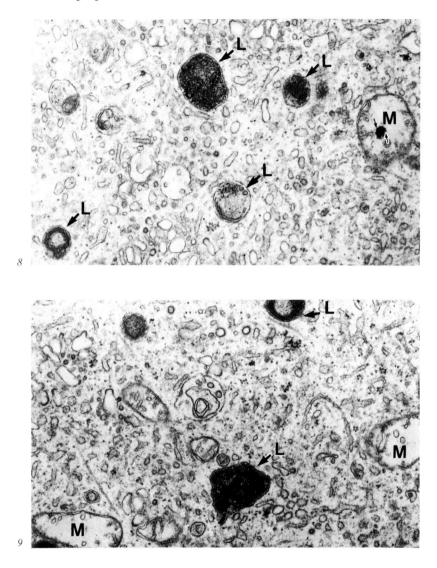

Fig. 5–9. Similar profiles of neurons from 38- and 48-day-old neuroblastoma cultures depicting similar degenerated mitochondria (M). Note the variation in form, density and number of lipofuscin granules (L). Not only is there an increased electron density of the pigment bodies, but more mitochondria appear to be converting into lipopigment (see electron-lucent bodies L, figures 5–8). In figure 9, the pigment (L) is deposited concentrically, while the one below it has denser areas within its structure. × 20,000.

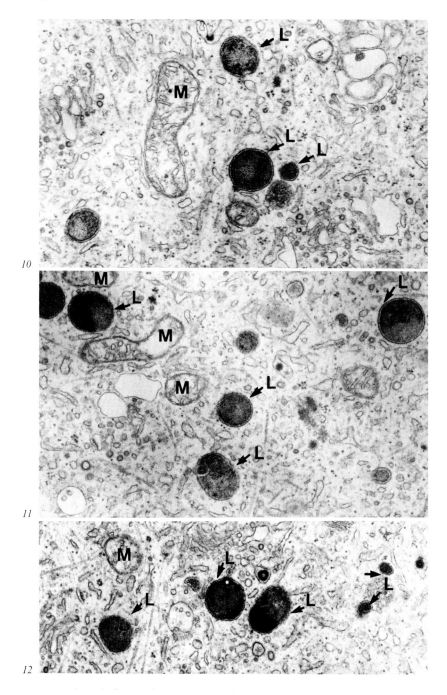

radical-induced peroxidation of unsaturated lipids is a major determinant of ageing [*Harman*, 1956, 1972; *Dormandy*, 1978; *Gordon*, 1974; *Miquel* et al., 1980]. *Silipandri* et al. [1979] provide evidence that the decline of the mitochondrial energy-dependent processes during ageing is due to the oxidation of pairs of neighboring thiol groups of the inner membrane, which may in turn induce changes in the permeability of the mitochondrial membrane. Thus old mitochondria may be more susceptible and thereafter more easily damaged [*Vanneste and van den Bosch de Aguilar*, 1981]. The increase in the average life span of mice produced by adding antioxidants to the diet may be attributed to a significant decrease in the level of deleterious free radical reactions throughout the body except in the mitochondria [*Harman*, 1961, 1968a, b]. The high rate of free radical initiation and the short chain lengths of free radical reactions in the mitochondria make it difficult to protect these organelles from free radical-induced reactions [*Harman*, 1972]. On the other hand one cannot exlude the fact that free radicals arising from oxidation-reduction reactions induce oxidation-polymerization processes involving the unsaturated mitochondrial lipids but also the mitochondrial DNA [*Harman*, 1972; *Miquel* et al., 1980; *Davies and Fotheringham*, 1981].

Concerning the origin of the intracellular ageing pigment, lipofuscin, many suggestions have been made as to the possible organelles involved in its formation, and this has been discussed [*Glees and Hasan*, 1976]. *Siakotos* et al. [1977] proposed a modified Novikoff sequence which involves the fusion of lipid globules or cytolysosomes with primary lysosomes and microperoxisomes. This structure eventually becomes the residual body of the autophagic sequence, termed the 'mature' lipofuscin granule. This schema is based on the sedimentation velocity isolation of lipofuscin from human brains with an age range of 0.36–78.8 years [*Siakotos* et al., 1970]. Up to approximately 17 years of age, the liposfuscin pigment is composed of electron-lucent globular areas at first, with an increasing electron-dense component added with time. A similar phenomenon is observed in the neuroblastoma cultures. The electron density of the pigment bodies increased with the age of the cultured neuroblastoma cells. We have modified the Gerl of

---

Fig. 10–12. Once more profiles of neurons from older neuroblastoma cultures (52-, 60- and 68-day-old cultures). The cells have numerous round to oval lipofuscin granules (L) with differences in the consistency of the matrix, some showing light and dark areas surrounded by a double-limiting membrane. Note the cristae-like structures within the osmiophilic bodies. Similarly there is a large number of degenerating or senescent mitochondria (M). × 20,000.

Novikoff et al. [1971] in our previous studies on age pigment in culture and have added mitochondria to the Gerl sequence [*Spoerri and Glees*, 1973, 1974]. Presently, our theory would involve the fusion of lipid globules formed within the mitochondria with primary lysosomes and microperoxisomes [*Novikoff* et al., 1973; *Siakotos and Munkres*, 1982]. The intramitochondrial presence of superoxide is of great significance to mechanisms of cell ageing. Hydrogen peroxide (H_2O_2) may inflict serious damage to the mitochondrial DNA and to the lipids and proteins of the inner membrane. Moreover, the superoxide radical and H_2O_2 have the capacity to generate other reactive oxygen species such as the hydroxyl radical and singlet molecular oxygen, which in turn, initiate radical chain reactions leading to extensive lipid and organic peroxide formation [*Tappel and Salkin*, 1959]. These chain reactions are detrimental to the respiratory activity of the mitochondria because a disruption of lipophilic sites by peroxidation simulates a detergent effect in the inactivation of enzyme systems [*Miquel* et al., 1980]. Further membrane disorganization may be induced by malonaldehyde, a major product of peroxidation of unsaturated lipids, which may react with primary amino groups and amino acids and proteins in a cross-linking reaction [*Siakotos and Munkres*, 1982].

The present study indicates that the accumulation of 'unusual' substances or structures appears to be a frequent characteristic of senescent mitochondria. The variability in the cristae or matrix is obvious, which can be the result of a metabolic disfunction. As the spreading alteration occurs only in senescent (30-day-old) neurons in culture and older, one could postulate that in younger cultures or animals [*Vanneste and van den Bosch de Aguilar*, 1981], cellular regulation mechanisms associated with a rapid turnover eliminate impaired mitochondria quickly. The very fact that accumulation of impaired mitochondria and lipofuscin pigment is associated and that many appositions are observed between them supports the evidence that mitochondria are directly involved in the genesis of lipofuscin.

References

Ahmed, M.M.; Glees, P.: Mitochondria degeneration after organic phosphate poisoning in prosimian primates. Cell Tissue Res. *175:* 459–465 (1976).

Bourne, G.H.: Lipofuscin. Neurological aspects of maturation and aging. Prog. Brain Res. *40:* 187–201 (1973).

Brizzee, K.R.; Kaack, B.; Klara, P.: Lipofuscin: intra- and extraneuronal accumulation and regional distribution; in Terry, Gerschon, Neurobiology of aging, vol. 3, pp. 229–244 (Raven Press, New York 1975).

Colcolough, H.L.; Hack, M.H.; Helmy, F.M.; Vaughn, G.E.; Veith, D.C.: Some histochemical, biochemical, morphological observations relating to lipofuscin and mitochondria. Acta histochem. *43:* 98–109 (1972).

Davies, I.; Fotheringham, A.P.: Lipofuscin – does it affect cellular performance. Exp. Gerontol. *16:* 119–125 (1981).

Dormandy, T.: Free-radical oxidations and antioxidants. Lancet *i:* 647–650 (1978).

Glees, P.; Gopinath, G.: Age changes in the centrally and peripherally located sensory neurons in rat. Z. Zellforsch. mikrosk. Anat. *141:* 285–289 (1973).

Glees, P.; Hasan, M.: Lipofuscin in neuronal aging and diseases; in Bargman, Doerr, Normale und pathalogische Anatomie, vol. 32, pp. 1–68 (Thieme, Stuttgart 1976).

Glees, P.; Hasan, M.; Spoerri, P.E.: Mitochondrial genesis of lipofuscin-evidence based on electron-microscopic studies of the brain, neural tissue culture and heart. J. Physiol., Lond. *239:* 87 (1974).

Glees, P.; Spoerri, P.E.; El Ghazzawi, E.: An ultrastructural study of hypothalamic neurons in monkeys of different ages with special reference to age related lipofuscin. J. Hirnforsch. *16:* 379–394 (1975).

Gordon, P.: Free radicals and the aging process; in Rockstein, Theoretical aspects of aging, pp. 61–81 (Academic Press, New York 1974).

Harman, D.: Aging: a theory based on free radical and radiation chemistry. J. Geront. *11:* 298–300 (1956).

Harman, D.: Prolongation of the normal life span and inhibition of spontaneous cancer by antioxidants. J. Geront. *16:* 247–254 (1961).

Harman, D.: Free radical theory of aging: effect of free radical reaction inhibitors on the mortality rate of male, LAF, mice. J. Geront. *23:* 476–482 (1968a).

Harman, D.: Free radical theory of aging: effect of free radical inhibitors on the life span of male, LAF, mice – second experiment. Gerontologist *8:* III, 13 (1968b).

Harman, D.: The biologic clock: the mitochondria? J. Am. Geriat. Soc. *20:* 145–147 (1972).

Hasan, M.; Glees, P.: Genesis and possible dissolution of neuronal lipofuscin. Gerontologia *18:* 217–236 (1972).

Hasan, M.; Glees, P.; Spoerri, P.E.: Dissolution and removal of neuronal lipofuscin following dimethylaminoethyl *p*-chlorophenoxyacetate administration to guinea pigs. Cell Tissue Res. *150:* 369–375 (1974).

Koobs, D.H.; Schultz, R.L.; Jutzy, R.V.: Origin of lipofuscin and possible consequences to myocardium. Archs Pathol. Lab. Med. *102:* 66–68 (1978).

Miquel, J.; Economos, A.C.; Fleming, J.; Johnson, J.E.: Mitochondrial role in cell aging. Exp. Gerontol. *15:* 575–591 (1980).

Novikoff, A.B.; Novikoff, P.M.; Quintana, N.; Davis, C.: Studies in microperoxisomes. IV. Interrelations of microperoxisomes, endoplasmic reticulum and lipofuscin granules. J. Histochem. Cytochem. *21:* 1010–1020 (1973).

Novikoff, P.M.; Novikoff, A.B.; Quintana, N.; Hauw, J.: Golgi apparatus, Gerl and lysosomes of neurons in rat dorsal root ganglia, studied by thick and thin cytochemistry. J. Cell Biol. *50:* 859–886 (1971).

Samorajsky, T.; Ordy, J.M.; Keefe, T.R.: The fine structure of lipofuscin age pigment in the nervous system of aged mice. J. Cell Biol. *26:* 779–795 (1965).

Siakotos, A.N.; Armstrong, D.; Koppang, N.; Muller, J.: Biochemical significance of age pigment in neurones; in Nandy, Sherwin, I. The aging brain and senile dementia, pp. 99–118 (Plenum Press, New York 1977).

Siakotos, A.N.; Munkres, K.D.: Recent developments in the isolation and properties of autofluorescent lipopigments; in Armstrong, Koppang, Rider, Ceroid-lipofuscinosis (Batten's disease), pp. 167–183 (Elsevier Biomedical Press, Amsterdam 1982).

Siakotos, A.N.; Watanabe, I.; Saito, A.; Fleischer, S.: Procedures for the isolation of two distinct lipopigments from human brain: lipofuscin and ceroid. Biochem. Med. *4:* 361–375 (1970).

Silipandri, N.; Silipandri, D.; Zoccarato, F.; Toninello, A.; Rugolo, M.: Changes in mammalian mitochondria during ageing. Bull. Mol. Biol. Med. *4:* 1–14 (1979).

Spoerri, P.E.: Ultrastructural age changes in neurons and myocardium in culture: the effects of centrophenoxine on age pigment; in Armstrong, Koppang, Rider, Ceroid-lipofuscinosis (Batten's disease), pp. 369–384 (Elsevier Biomedical Press, Amsterdam 1982).

Spoerri, P.E.; Dresp, W.; Heyder, E.: A simple embedding technique for monolayer neuronal cultures grown in plastic flasks. Acta anat. *107:* 221–223 (1980a).

Spoerri, P.E.; Glees, P.: Neuronal aging in cultures; an electron-microscopic study. Exp. Gerontol. *8:* 259–263 (1973).

Spoerri, P.E.; Glees, P.: The effects of dimethylaminoethyl p-chlorophenoxyacetate on spinal ganglia neurons and satellite cells in culture. Mitochondrial changes in the ageing neurons. An electron microscope study. Mech. Age. Dev. *3:* 131–135 (1974).

Spoerri, P.E.; Glees, P.: Ultrastructural reactions of spinal ganglia to trio-ortho-cresylphosphate. Effects of neurotoxicity. Cell Tissue Res. *199:* 409–414 (1979).

Spoerri, P.E.; Glees, P.; Dresp, W.: The time course of synapse formation of mouse neuroblastoma cells in monolayer culture. Cell Tissue Res. *205:* 411–421 (1980b).

Spoerri, P.E.; Glees, P.; El Ghazzawi, E.: Accumulation of lipofuscin in the myocardium of senile guinea pigs: dissolution and removal of lipofuscin following dimethylaminoethyl p-chlorophenoxyacetate administration. Mech. Age. Dev. *3:* 311–321 (1974).

Strehler, B.L.: On the histochemistry and ultrastructure of age pigment; in Advances in gerontological research, vol. 1, pp. 257–288 (Academic Press, New York 1964).

Strehler, B.L.; Mark, D.; Mildvan, A.S.: Rate and magnitude of age pigment accumulation in the human myocardium. J. Geront. *14:* 430–439 (1959).

Tappel, A.L.; Salkin, H.: Lipid peroxidation in isolated mitochondria. Archs Biochem. Biophys. *80:* 326–330 (1959).

Tauchi, H.; Sato, T.: Age changes in size and number of mitochondria of human hepatic cells. J. Geront. *23:* 509–512 (1968).

Tomanek, R.J.; Karlson, U.L.: Myocardial ultrastructure of young and senescent rats. J. Ultrastruct. Res. *40:* 201–220 (1973).

Travis, D.F.; Travis, A.: Ultrastructural changes in the left ventricular rat myocardial cells with age. J. Ultrastruct. Res. *39:* 124–148 (1972).

Vanneste, J.; van den Bosch de Aguilar, P.: Mitochondrial alterations in the spinal ganglion neurons in ageing rats. Acta neuropath. *54:* 83–87 (1981).

Dr. P.E. Spoerri, Department of Anatomy, Georg-August-Universität,
Kreuzbergring 36, D-3400 Göttingen (FRG)

Evidence that Paromomycin Induces Premature Ageing in Human Fibroblasts

R. Holliday, S.I.S. Rattan[1]

National Institute for Medical Research, London, UK

Introduction

The aminoglycoside antibiotic, paromomycin (Pm), has been shown to reduce the fidelity of translation by eukaryotic ribosomes in vitro [1, 14, 20, 21]. It can also bring about the phenotypic suppression of mutations in yeast [10, 11, 15, 19], *Podospora* [3] and Chinese hamster cells [6]. It has been shown that Pm is taken up by cultured human fibroblasts and that it is associated with the lysosomal fraction [2, 16]. If it also binds to ribosomes and introduces ambiguity in messenger translation, it provides a powerful tool to test the protein error theory of cellular ageing. This proposes that the intrinsic changes which eventually bring about the senescence and death of fibroblast populations are due to a progressive feedback of errors in the machinery of protein synthesis [12, 13]. A prediction of the theory is that error-promoting agents should accelerate the process of ageing. We present evidence here that Pm has this effect on populations of fibroblasts.

Materials and Methods

Cell Cultures
All experiments were carried out with human fetal lung fibroblast strain MRC-5, which was originally characterised by *Jacobs* et al. [9] and has a longevity of 50–70 population doublings [5]. For life span experiments, cells were grown in 25 cm^2 Falcon flasks or 1.75 cm^2 wells

---

[1] The studies on autofluorescence would not have been possible without Mr. *Keith Keeler's* skilled assistance. We also thank Dr. *J.H. Buchanan* for his interest and helpful discussion.

using minimal essential medium F-15 (Gibco Biocult), supplemented with 10% fetal calf serum and antibiotics (100 U penicillin and 100 µg streptomycin per ml). Paromomycin sulfate was kindly supplied by Parke-Davis & Co., Pontypool, Gwent, Wales. A stock solution of 10 mg/ml in F-15 medium was diluted as required. Confluent cultures were rinsed with phosphate-buffered saline and harvested with trypsin-versene. Cells were counted immediately after trypsinisation using a Coulter counter, to calculate cumulative population doublings (CPD). For serial passaging, cultures were split 1:4 or 1:8 when grown vigorously, and 1:2 and 1:4 when growth slowed down. Cultures were terminated when they failed to become confluent after at least 4 weekly medium changes. Growth rate was determined by inoculating flasks with a constant number of cells. Duplicate flasks were trypsinised and counted after 6 h and at daily intervals thereafter.

Glucose-6-Phosphate Dehydrogenase

Cells were grown in 81 cm^2 Falcon flasks and routinely split 1:2 or 1:4. To retain equivalence with population doublings, these were scored as 1 and 2 passages, respectively. As soon as cultures became confluent, the cells were harvested by trypsinisation from one or two flasks, washed with a mixture of culture medium and buffer (0.005 mM Tris/HCl in 0.15 M NaCl) to inactivate the trypsin, and then rewashed with buffer. The pellet was resuspended in 0.5 or 1 ml of extraction buffer, as previously described [7]. The cells were disrupted by sonication and cell debris removed by spinning at 10,000 rpm for 30 min at 4°C. The assay mixture was the same as previously described [7]. For each assay 0.02 ml enzyme was added to and mixed with 0.78 ml assay mixture and the rate of increase in OD at 340 nm was measured over a 1- to 1.5-min period using a Unicam SP800 UV spectrophotometer. The enzyme activity prior to heating was in all cases the average of three separate determinations. Approximately 0.4 ml of cell extract was placed at zero time in a Grant waterbath accurate to within 0.1°C. The inactivation temperature was between 59 and 60°C, usually 59.5°C was chosen. Samples were removed at 2- to 4-min intervals and immediately assayed at 25°C during a 20- to 35-min period of heat inactivation. Repeated use of this procedure has shown that the estimate of the proportion of heat-labile enzyme is reproducible to within 2-3%.

Autofluorescence

Trypsinised cells were analysed for autofluorescence (AF) and size using a fluorescence-activated cell sorter (FACS-11; Beckton & Dickenson) equipped with an argon ion laser (Spectraphysics 164-05, UV-enhanced, set at 351 nm). For each determination, 10^4 cells were measured using the procedures previously described [17].

Results

Effects of Pm on Growth and Longevity

In initial experiments, cells were grown in a variety of concentrations of Pm in Falcon flasks or wells and it was shown that they achieved confluence as rapidly as control populations, with a similar cell yield, even in concentrations as high as 1 mg/ml. Characteristic growth curves are shown in figure 1. When cells were subcultured in the presence of 1 mg/ml Pm, the normal

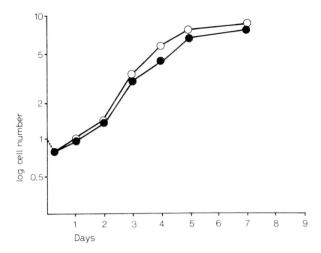

Fig. 1. Single-step growth curves of MRC-5 fibroblasts in normal medium (closed circles) and medium containing 1 mg/ml Pm (open circles).

growth rate was maintained for many generations, although, as we shall show, immediate effects on these cells do occur. In long-term experiments, when cells were maintained in this concentration of Pm, the growth rate eventually decreased and the cultures always ceased growth prior to the untreated controls. The extent of this effect is, in part, dependent on the culture age when Pm is first added. In general, cells become more sensitive to the antibiotic in terms of reduced growth potential and later the passage level at which treatment begins. This is documented in table I and figure 2. In different experiments we take the control life span in CPD as 100%; we then calculate the percent of the life span which has been completed when Pm is added and we calculate the percentage of the remaining life span (i.e. that of the controls), which will be completed in the presence of Pm. This makes it possible to make quantitative comparisons between different longevity experiments, since it is well known that untreated cultures of MRC-5 vary considerably in the number of CPD completed before growth ceases [5]. The results of several experiments are shown in figure 2. The average reduction in the remaining life span for young cultures is about 35%, for middle-aged cultures it is 45–55% and for old cultures it is about 65%.

Experiments were also carried out with lower concentrations of Pm, in the range 100–500 μg/ml. In most cases, the longevity in CPD was reduced, but differences from controls were not always significant. We did observe,

Table I. The effect of Pm on the longevity (CPD) of MRC-5 cultures

Experiment No.		Growth in CPD		Final life span (CPD)
		prior to Pm	in Pm	
1	Control	16^1 + 42.9	–	58.9
	Pm treated	16	26.6	42.6
		24.3	19.2	43.5
		35.6	8.4	44.0
2	Control	15^1 + 49.2	–	64.2
	Pm treated	15	24.5	39.5
		24.4	24.1	48.5
		45.5	8.1	52.6
		58.9	1.8	60.7
3	Control	17.5^1 + 50.6	–	68.1
	Pm treated	17.5^2	36.0, 35.5, 35.5, 37.0	53.5 (mean)
		26.2	28.6	54.8
		43.3	14.1	57.4
		51.5	3.7	55.2
		61.7	3.8	65.5
		65.2	1.9	67.1

Pm was added at the CPD level indicated (first column) and the cultures grown in the presence of the antibiotic until proliferation ceased.
[1] Passage level of initial culture.
[2] 4 cultures were treated with Pm.

however, that the cells were more sensitive to these concentrations of the antibiotic in terms of CPD achieved, when they were grown in 1.75 cm² wells rather than in flasks (results not shown). It is known that population size can affect life span [5], and this may also be the case for the experiments with Pm.

We also examined the effect of removing Pm from the medium after growth had slowed down in the presence of the antibiotic. In experiment 2 (table I), passage 15 cells were treated with Pm for 20.9 CPD (62 days) and then returned to normal medium. They grew a further 11.2 CPD before growth ceased, which was 17 CPD less than the life span of the control cultures. In experiment 3 (table I), cells were removed from Pm at three different times. These results are shown in figure 3, together with growth curves of treated and untreated populations. It can be seen that limited recovery occurs, but the cells do not achieve the growth rate of controls at the equivalent CPD level or at the equivalent time in days; nor do they recover to a

Ageing Induced by Paromomycin 225

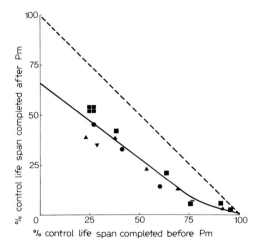

Fig. 2. The reduction of life span by Pm in different experiments. The diagonal dashed line represents the control cultures, all of which achieve 100% life span. The symbols and solid line show the % reduction in the longevity of cultures transferred to medium containing Pm at different times during their life span. Exp. 1, table I, circles; exp. 2, table I, triangles; exp. 3, table I, squares; two other experiments, inverted triangles.

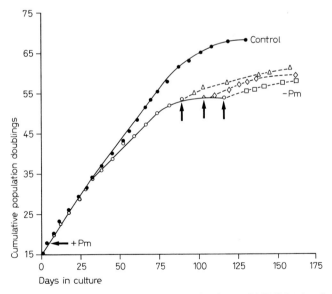

Fig. 3. Cumulative growth of an untreated culture of MRC-5 (closed circles) and one grown from CPD 17.5 in medium containing 1 mg/ml Pm (open circles). Arrows indicate the times Pm was removed, and the broken line the cultures' subsequent growth.

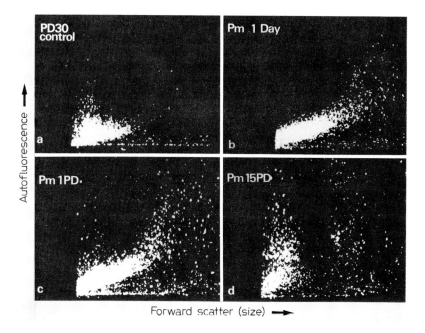

Fig. 4. The distribution patterns (dot plots) of autofluorescence (AF) and size in Pm treated cells recorded by a FACS II. *a* Control MRC-5 cells. *b* MRC-5 cells grown in 1 mg/ml Pm for 1 day; *c, d* for 1 CPD (5 days) and 15 CPD, respectively. Each dot is the signal from one cell.

full life span. These results show there is a 'carry-over' effect from Pm treatment.

The reduced growth potential of MRC-5 cells in the presence of Pm is not due to a reduction in cell viability. The growth curves in figure 1 and the first 30–35 days in the experiment in figure 3 would not be identical if an increased proportion of Pm-treated cells was not dividing. We have also examined the percent re-attachment of cells after trypsinisation and found no difference between Pm-treated and control populations.

Cellular Morphology

Cells grown in the presence of Pm increase in size (fig. 4) and become more granular when examined by phase-contrast microscopy. When the growth slows down the cells clearly resemble senescent, untreated cells. They do not line up in parallel arrays to form the characteristic whorls of confluent

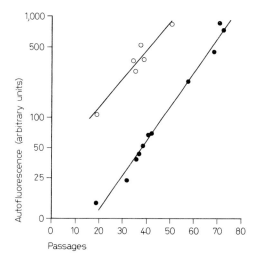

Fig. 5. The increase in autofluorescence (AF) during the serial subculture of untreated MRC-5 cells (closed circles) and in cells grown in the presence of 1 mg/ml Pm from passage 17 (open circles). The lines drawn are the best fit after regression analysis.

young cells; cell density at confluence declines; cells are highly granular, often with more than one nucleus, and cell debris accumulates in the medium.

Autofluorescence

We have previously shown that the AF of MRC-5, as measured by FACS, increases exponentially with culture age, which provides a valuable index of cell ageing [17]. We have examined the AF of cells grown for increasing periods of time in Pm. There is an initial rapid increase in AF over a 24-hour period of growth (fig. 4), followed by an exponential increase which runs parallel to the control cultures (fig. 5). The cultures which died prematurely in Pm have an amount of AF which is very comparable to the amount in late passage control cells. This provides further evidence that Pm is in fact speeding up the ageing process.

Glucose-6-Phosphate Dehydrogenase

In initial experiments, early passage cells were grown in 81 cm^2 Falcon flasks for 2–4 population doublings in medium containing either 250 or 500 µg/ml Pm. Cell-free extracts were prepared and the heat inactivation of

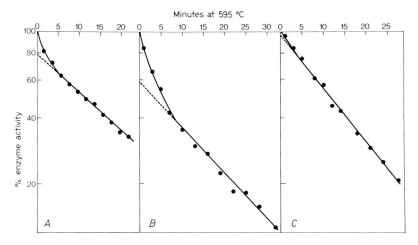

Fig. 6. The heat-lability of G-6-PD. *A* MRC-5 cells grown for two population doublings (1:4 split) in 500 µg/ml Pm. *B* The same culture grown in 500 µg/ml Pm to senescence at passage 51 (see also table II). *C* Untreated cells at passage 44.

G-6-PD was examined. A representative result is shown in figure 6a. Early passage MRC-5 cells contain 0–10% heat-labile G-6-PD and late passage cultures contain 15–25% [7]. In various experiments it was found that early passage cells grown in Pm contained 20–40% heat-labile G-6-PD.

In a long-term experiment, MRC-5 cells were grown continually in 500 µg/ml Pm in 81 cm^2 Falcon flasks and at every few passages, the stability of G-6-PD was examined. An initial effect on G-6-PD was seen, but as table II shows, most of the heat-labile G-6-PD disappeared on subsequent subculture in the presence of the antibiotic. Only when the culture became clearly senescent at passage 50 did a substantial proportion of heat-labile G-6-PD reappear (fig. 6b) and growth subsequently ceased at passage 55. The proportion of heat-labile G-6-PD in untreated cultures (table II, fig. 6c) was very much the same as reported previously [7]. The results strongly indicate that MRC-5 cells can respond to the presence of Pm, at least at this concentration, and in some way either prevent the formation or, perhaps more likely, effect the removal of altered G-6-PD. This ability to 'adapt' to Pm is dose-dependent, since a similar long-term experiment carried out with 1 mg/ml Pm showed no evidence of adaptation. In this case, a significant proportion of heat-labile G-6-PD was formed initially and which then fluctuated within the range 18–27% until growth ceased. These cells died out at passage 47, which was about 70% of the life span of the control population.

Table II. The percentage of heat-labile G-6-PD in control and Pm-treated cultures of MRC-5

Experiment No.	Controls (passage)	%heat-labile G-6-PD	Pm treatment	Passages in Pm	% heat-labile G-6-PD
1	20	9	500 μg/ml from passage 18	2	22
	24	8		6	23
	34	9		7	15
	39	12		10	16
	45	5		14	9
	56	13		19	13
	61	31		24	12
				33	41
				36	27
				37[1]	35
2	–	–	1 mg/ml from passage 15	5	30
				12	21
				15	30
				20	18
				28	22
				32[2]	27

[1] Died at passage 55 (control at passage 62).
[2] Died at passage 47 (control at passage 64).

Discussion

We have shown that quite high concentrations of Pm have no effect on growth rate of MRC-5 fibroblasts. However, on further sub-culture, the growth rate slows down and the cells resemble in appearance late passage, untreated cells. The relative sensitivity of MRC-5 cells appear to increase with culture age. Untreated senescent cells with a few CPD of growth remaining, scarcely divide at all in a concentration of Pm which has no initial effect on the growth of early passage cells. There thus appears to be a synergistic interaction between normal ageing and the effect of Pm. Moreover, cells which appear to have become prematurely senescent in the presence of Pm, show only limited recovery when they are returned to normal medium. Thus, Pm induces a physiological carry-over effect on growth and longevity.

There are also immediate effects of Pm on MRC-5 fibroblasts. We previously showed that the antibiotic is rapidly taken up by cells and concentrated in the lysosomal fraction; on returning the cells to normal medium, much of the Pm leaves the cell, but a proportion is retained [2, 16]. There is also a rapid increase in AF over a period of 24–48 h. Also, cells grown for at least two generations in the presence of Pm, contain a significantly increased fraction of heat-labile G-6-PD. The evidence that this altered G-6-PD is the result of errors in trancription or translation, rather than post-synthetic modifications, has been reviewed elsewhere [8]. However, we have so far no direct evidence that Pm binds to endoplasmic reticulum or ribosomes.

The results show that MRC-5 cells are able to respond or adapt to the presence of Pm. In the AF studies, the initial rapid increase did not continue beyond 48 h. Instead, the rise in AF subsequently ran parallel to that in control cultures (fig. 5). When cells were grown in 0.5 mg/ml Pm, the initial increase in heat-labile G-6-PD was followed by a fall to the control value (table II). We have previously seen the same effect when MRC-5 cells were grown in the presence of *p*-fluorophenylalanine [*Holliday and Tarrant*, unpublished results], and it is possible that cells respond to protein error-inducing agents or analogues by inducing scavenging proteases. With higher levels of Pm in the medium there may then be a balance between the formation of altered molecules and their removal, as shown by the fairly steady increased level of heat-labile G-6-PD (table II). It is striking that in all cases of fibroblast senescence, either late passage MRC-5, or MRC-5 treated with error-inducing agents, or in cells from premature ageing syndromes, the fraction of heat-labile G-6-PD is in the range 15–40% [7, 8].

The quantitative effects of Pm on longevity can be interpreted in two ways as illustrated in figure 7. In this figure, we assume that there is an exponential increase in some biochemical parameter X (which runs parallel to an increase in AF), until finally a lethal level of X is reached. The immediate effect of Pm may be to 'jump' or shift the cells to an increased physiological age; thereafter, the ageing process proceeds as before (line A). The lethal level of X is slightly lower in the presence of Pm (L^{Pm}) than in its absence (L^c), and when it is removed there is some limited recovery (line D). A second possibility is that the ageing process is speeded up; in other words, the rate of acceleration of X is simply increased and a lethal level is therefore reached more quickly (line B). The first interpretation would predict that the solid line in figure 2 should run parallel to the dashed control line and intersect with the abscissa 15–20 CPD in advance of 100% life span, whereas the second interpretation predicts that the solid line should gradually converge to the control and reach the abcissa at 100% lifespan. The observed results are intermediate between these two predictions. The two interpretations could

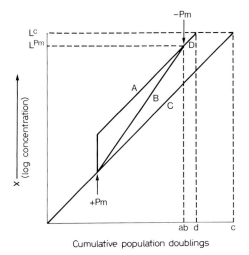

Fig. 7. Possible interpretations of the life-shortening effects of Pm. It is assumed that in vitro ageing is due to the exponential increase of a biochemical parameter X. Untreated cells (line C) cease growth after c population doublings (CPD). The addition of Pm may produce an initial rise in X followed by a slower exponential increase (line A). The lethal level of X is lower in the presence of Pm (L^{Pm}) than in its absence (L^c). Cells grown in the presence of Pm die after ab CPD, but if Pm is withdrawn limited recovery occurs (line D, and see fig. 3), until death after d CPD. An alternative interpretation is that the rate of accumulation of X is simply accelerated in the presence of Pm (line B).

also be distinguished by further experiments in which Pm is added to early passage cultures for a small number of CPD and then withdrawn. Figure 7 shows that a pulse treatment of Pm given to young cells would have little effect on life span if line B is correct, whereas with A a marked effect would be seen.

The error theory predicts that defects in protein synthesis will accumulate exponentially with time or CPD; errors could therefore be equated with X in figure 7. Recent claims that protein errors do not increase during the ageing of cultured human fibroblasts are based on inadequate methods, and there is indeed considerable indirect evidence in favour of the theory [for a review, see 4]. Evidence has been obtained that the aminoglycoside antibiotic, streptomycin, can induce a lethal protein 'error catastrophe' in a growing population of *Escherichia coli* cells [18]. It is not unreasonable to suppose that the related aminoglycoside, Pm, which is known to induce misreading of messenger RNA by eukaryotic ribosomes, would accelerate the normal process of cellular ageing, if this is indeed due to the accumulation of errors in proteins. The results obtained therefore provide further support for the error theory.

References

1   Buchanan, J.H.; Bunn, C.L.; Lappin, R.I.; Stevens, A.: Accuracy of in vitro protein synthesis: translation of polyuridylic acid by cell free extracts of human fibroblasts. Mech. Age. Dev. *12:* 339–353 (1980).
2   Buchanan, J.H.; Rattan, S.I.S.; Stevens, A.; Holliday, R.: Intracellular accumulation of a fluorescent derivative of paromomycin by human fibroblasts in culture. J. cell. Biochem. *20:* 71–80 (1982).
3   Coppin-Raynal, E.: Ribosomal suppressors and antisuppresors in *Podospora anserina:* altered susceptibility to paromomycin and relationships between genetic and phenotypic suppression. Biochem. Genet. *19:* 729–740 (1981).
4   Holliday, R.: The unsolved problem of cellular ageing; in Sauer, Cellular ageing; Monogr. devl Biol. 17, pp. 60–77 (Karger, Basel 1984).
5   Holliday, R.; Huschtscha, L.I.; Tarrant, G.M.; Kirkwood, T.B.L.: Testing the commitment theory of cellular ageing. Science *198:* 366–372 (1977).
6   Holliday, R.; Jeggo, P.A.: Phenotypic suppression of mutants of Chinese hamster ovary cells by paromomycin (in preparation).
7   Holliday, R.; Tarrant, G.M.: Altered enzymes in ageing human fibroblasts. Nature, Lond. *238:* 26–30 (1972).
8   Holliday, R.; Thompson, K.V.A.: Genetic effects on the longevity of cultured human fibroblasts. III. Correlations with altered glucose-6-phosphate dehydrogenase. Gerontology *29:* 89–96 (1983).
9   Jacobs, J.P.; Jones, C.M.; Baillie, J.P.: Characteristics of a human diploid cell designated MRC-5. Nature, Lond. *227:* 168–170 (1970).
10 Musurekar, M.; Palmer, E.; Ono, B.I.; Wilhelm, J.M.; Sherman, F.: Misreading of the ribosomal suppressor SUP46 due to an altered 40S subunit in yeast. J. molec. Biol. *147:* 381–390 (1981).
11 Mironova, L.M.; Provorov, N.A.; Ter-Avanesyan, M.D.; Inge-Vechtomov, S.G.; Smirnov, V.N.; Surguchov, A.P.: The effects of paromomycin on the expression of ribosomal suppressors in yeast. Curr. Genet. *5:* 149–152 (1982).
12 Orgel, L.E.: The maintenance of the accuracy of protein synthesis and its relevance to ageing. Proc. natn. Acad. Sci. USA *49:* 517–521 (1963).
13 Orgel, L.E.: Ageing of clones of mammalian cells. Nature, Lond. *243:* 441–445 (1973).
14 Palmer, E.; Wilhelm, J.M.: Mistranslation in a eukaryotic organism. Cell *13:* 329–334 (1978).
15 Palmer, E.; Wilhelm, J.M.; Sherman, F.: Phenotypic suppression of nonsense mutants in yeast by aminoglycoside antibiotics. Nature, Lond. *277:* 148–149 (1979).
16 Rattan, S.I.S.: Uptake, accumulation and release of paromomycin by human fibroblasts in culture. IRCS med. Sci. *10:* 807–808 (1982).
17 Rattan, S.I.S.; Keeler, K.D.; Buchanan, J.H.; Holliday, R.: Autofluorescence as an index of ageing in human fibroblasts in culture. Biosci. Rep. *2:* 561–567 (1982).

18   Rosenberger, R.F.: Streptomycin-induced protein error propagation appears to lead to cell death in *Escherichia coli.* IRCS med. Sci. *10:* 874–875 (1982).
19   Singh, A.; Ursic, D.; Davies, J.: Phenotypic suppression and misreading in *Saccharomyces cerevisiae.* Nature, Lond. *277:* 146–148 (1979).
20   Wilhelm, J.P.; Pettitt, S.E.; Jessop, J.J.: Aminoglycoside antibiotics and eukaryotic protein synthesis: structure-function relationship in the stimulation of misreading with a wheat embryo system. Biochemistry *17:* 1143–1149 (1978).
21   Wilhelm, J.M.; Jessop, J.J.; Pettitt, S.E.: Aminoglycoside antibiotics and eukaryotic protein synthesis. Stimulation of errors in the translation of natural messengers in extracts of cultured human cells. Biochemistry *17:* 1149–1153 (1978b).

R. Holliday, PhD, FRS, National Institute for Medical Research, The Ridgeway, Mill Hill, London NW7 1AA (England)

Effect of in vitro Ageing on the Transport of Neutral Amino Acids in Human Fibroblasts[1]

Gian C. Gazzola, Ovidio Bussolati, Nicola Longo, Valeria Dall'Asta, Renata Franchi-Gazzola, Guido G. Guidotti

Istituto di Patologia Generale, Università di Parma, Italy

Functional changes in cell membranes have been reported to play a key role in ageing [*Grinna*, 1977; *Zs.-Nagy*, 1979]. They include alterations in the transport of ions [*Turk and Milo*, 1974], nutrients [*Polgar* et al., 1978; *Wheeler*, 1981, 1982] and proteins [*Berumen and Macieira-Coelho*, 1977]. Ageing at the cell level has been widely studied using cultured human fibroblasts [*Hayflick*, 1980]. With this biological model, some of the features concerning age-related changes in the transport of sugars [*Germinario* et al., 1980; *Cremer* et al., 1981] and nucleosides [*Polgar* et al., 1978] have been elucidated. The effect of ageing on the transport of amino acids in these cells has been studied less extensively [*Chen* et al., 1980; *Goldstein* et al., 1976; *Hill*, 1977] and no attempts have been made to identify the transport system(s) involved. Recently, we have characterized the inward transport of neutral amino acids in cultured human fibroblasts [*Gazzola* et al., 1980; *Franchi-Gazzola* et al., 1982] and defined several regulatory mechanisms acting thereon [*Gazzola* et al., 1980, 1981b]. Therefore, we undertook this investigation with two aims: (a) to explore the effect of in vitro ageing on neutral amino acid uptake by human fibroblasts, and (b) to identify the transport systems possibly affected by the ageing process.

Materials and Methods

Cell Culture

Skin biopsy explants from a 16-year-old female donor without any known disease were the source of human fibroblasts primary cultures [*Gazzola* et al., 1980]. The initial amplification

---

[1] This work was supported by the CNR Gruppo Nazionale Struttura e Funzione di Macromolecole Biologiche and by the Ministero della Pubblica Istruzione Gruppo Biologia e Patologia delle Membrane, Rome, Italy.

required four passages; the cells were then stored in liquid nitrogen. After rapid thawing, the content of ampoules (1.5×10^6 cells) was plated onto 10-cm diameter plastic dishes (Costar) and the cells were grown to confluence in Medium 199 containing 10% fetal calf serum, penicillin (100 units/ml) and streptomycin (100 µg/ml). Conditions of culturing were: pH 7.4; atmosphere 5% CO_2 in air; temperature 37°C.

Ageing of the cell population was pursued by the following split procedure. Upon reaching confluence, the cells were harvested by trypsinization [Gazzola et al., 1980] and counted using a Microcellcounter (Toa Electric Co., type CC-1006): 4×10^5 cells were plated under the conditions described above and newly grown to confluence; the remaining cells were seeded (3×10^4 cells/cm^2) onto 2-cm^2 wells of disposable multiwell trays (Costar) and used for transport assay after selected periods of culturing. In long-lasting cell cultures the growth medium was renewed every 72 h (and always 22 h before uptake assay). The split process was sequentially repeated for the period covering the entire lifespan of the cell population. In this interval, the number of population doublings was determined within each passage using the expression [Bemiller and Miller, 1979]:

$$\ln(B/A)/\ln 2,$$

where B is the number of cells at confluence and A is the number of cells attached to the substratum after their inoculation into the dish. The latter number has been calculated on the basis of a settling efficiency of 0.8 at 22 h after seeding (a period in which no cell proliferation takes place) [Macieira-Coelho et al., 1966]. This efficiency did not change appreciably during the lifespan of the cell population.

Transport Assay

Human fibroblast monolayers grown for 22 h (non-confluent cultures) or for 7–14 days (confluent cultures) in 2-cm^2 wells containing 1 ml of complete 199 Medium (see above) were preincubated for 90 min in Earle's balanced salt solution containing 10% dialyzed fetal calf serum [Gazzola et al., 1980]. The amino acid-depleted cells were then washed and incubated for 1 min at 37°C in 0.2 ml of Earle's balanced salt solution containing the ^{3}H-labelled amino acid at 0.1 mM (L-proline, L-leucine) or 0.05 mM (L-serine) final concentration. In designated experiments, a Na^+-free medium (in which choline replaced Na^+ in the sodium salts of the Earle's mixture) was used. Amino acid uptake was assessed using the cluster-tray method for rapid measurement of solute fluxes in adherent cells described by Gazzola et al. [1981a]. Acid-soluble pools were extracted with 10% trichloroacetic acid and counted in a liquid scintillation spectrometer. The cells were dissolved in 1 N NaOH and assayed for protein directly in the wells using a modified Lowry procedure [Wang and Smith, 1975], as described previously [Gazzola et al., 1981a]. The fibroblast monolayers from partner wells were trypsinized and the resulting cells were suspended and counted by an electronic cell counter.

Calculations

Amino acid uptake was expressed either as nanomoles per milligram of cell protein per minute, or as femtomoles per cell per minute. The population doubling level (PDL) reached by cultured human fibroblasts at the time of each uptake measurement was calculated by adding the cumulative number of doublings within each passage after preliminary amplification to the number of doublings occurred during the incubation of monolayers in multiwell trays; the latter term was null for non-confluent cultures (incubated for 22 h).

Materials

All sera, growth media, antibiotics and trypsin solution were from Gibco, N.Y. L-[5-^{3}H]-Proline (29 Ci/mmol), L-[4,5-^{3}H]-leucine (48 Ci/mmol) and L-[3-^{3}H]-serine (14 Ci/mmol) were obtained from Amersham, Bucks, England. Sigma was the source of all unlabeled amino acids and other chemicals.

Results

In vitro Senescence of Skin-Derived Human Fibroblasts

The data shown in figure 1 represent changes that occur during the in vitro ageing of the fibroblast population within the period encompassed by our experiments (see 'Cell Culture'). The cell strain characteristically underwent an increased doubling time (36–120 h, fig. 1a) and a decreased cell density attained at confluence (1.2×10^5 to 2.5×10^4 cells/cm^2, fig. 1b) with increasing age in vitro.

The cells in culture ceased to replicate after about 35 population doublings. At this time, the cells appeared large and granular and no mitotic figures were observed. The cell strain was considered senescent (Hayflick's phase III, [*Hayflick and Moorhead*, 1961]) when the split process yielded cell subcultures that failed to show a substantial growth in three weeks.

Amino Acid Transport and Ageing

Three main systems of mediation, the Na$^+$-dependent systems ASC and A and the Na$^+$-independent system L, participate to the transport of neutral amino acids in human fibroblasts [*Gazzola* et al., 1980; *Franchi-Gazzola* et al., 1982]. Systems ASC and A exhibit reactivity toward amino acids with polar, linear side chains; system A distinctively extends it to the imino acid L-proline and accepts N-methylation of the substrate. System L shows a preference for amino acids with apolar, bulky side chain [*Gazzola* et al., 1980]. The activity of these systems has been assessed by measurements of the uptake (initial entry rates) of L-serine (for system ASC), L-proline (for system A) and L-leucine (for system L) under experimental conditions of transport assay that discriminate among these systems [*Gazzola* et al., 1980; *Franchi-Gazzola* et al., 1982], following a proper depletion of the intracellular amino acid pools (see 'Transport Assay').

Figure 2 shows that the activity of these three systems was remarkably lower when measured in phase-III senescent human fibroblasts than in cells beginning phase II (first passage after the initial amplification). These differences were more marked with data expressed as nanomoles per milligram of

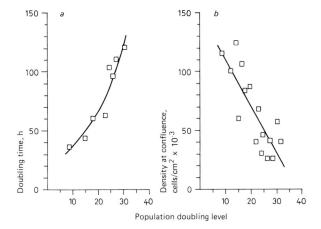

Fig. 1. Doubling time *(a)* and density at confluence *(b)* of human cultured fibroblasts at increasing population doubling level. Data obtained as described in the text. Lines are computer-derived best fits to a single exponential equation *(a)* and to a linear regression equation *(b)*.

protein per minute (fig. 2b) than as femtomoles per cell per minute (fig. 2a), though always statistically significant ($p < 0.005$). This might be due to a variation in cell protein content during ageing of cell population [*Wang* et al., 1970].

The rate of change in activity of the Na^+-dependent systems ASC and A during in vitro ageing (expressed as PDL) of phase-II human fibroblasts is shown in figure 3. When non-confluent cultures were used the initial uptake of *L*-serine (System ASC, fig. 3a) and of *L*-proline (system A, fig. 3c) decreased significantly with ageing of the population. The regression coefficient was lower with transport activity expressed on a 'per cell' than on a 'per milligram protein' basis. A significant decrease of transport activity with ageing was also observed with confluent fibroblast cultures when *L*-serine was the tracer amino acid (fig. 3b). Conversely, under the same culture conditions, no significant age-related decrement of *L*-proline uptake was detected (fig. 3d).

As shown in figure 1, the fibroblast cultures reached confluence at cell densities that decreased as the age of the population increased. On the other hand, amino acid transport by systems A and ASC is known to change markedly with cell density in fibroblast-like cell culture [*Guidotti* et al., 1978; *Borghetti* et al., 1980; *Piedimonte* et al., 1982]. Therefore, an attempt was made

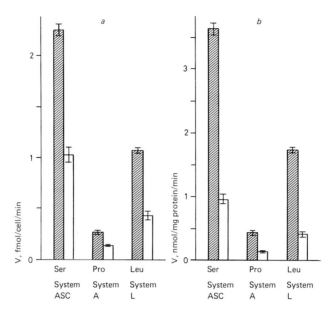

Fig. 2. Amino acid uptake by phase-II (▨) and phase-III (☐) cultured human fibroblasts. Confluent cultures were depleted of intracellular amino acids, washed and incubated for 1 min at 37 °C in a Na^+-containing medium (Earle's balanced salt solution) supplemented with 0.1 mM L-proline or 0.05 mM L-serine and in a Na^+-free medium in which choline replaced the cation in the sodium salts of the Earle's mixture, supplemented with 0.1 mM L-leucine. The results are shown as means of triplicate determinations with the standard deviation.

to evaluate the relative contribution of culture density to the age-dependent changes in transport activity of systems ASC and A. Multiple regression analyses in which amino acid transport was the dependent variable and ageing (PDL) and cell density were the independent variables have been applied to all the data reported as femtomoles per cell per minute in figure 3 for non-confluent and confluent cultures. This analysis for L-serine uptake (system ASC) showed a significant regression ($R^2 = 0.51$; $F = 7.7$; $p < 0.01$) graphically represented as a surface in figure 4. Most of the deviance given by the regression arose from ageing ($F = 15.3$; $p < 0.01$), whereas the contribution of cell density was not significant ($F = 1.2$). Therefore, the in vitro ageing of the population accounted almost entirely for the rate of change in transport activity by system ASC. Figure 5 shows the graphical representation of the surface obtained from the multiple regression for L-proline uptake (system A). Also in this case the regression was significant ($R^2 = 0.65$; $F = 14.2$;

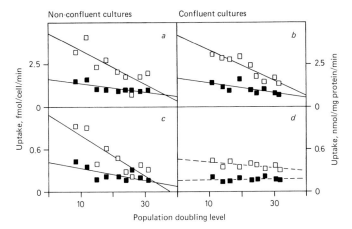

Fig. 3. L-Serine uptake (system ASC; *a* and *b*) and L-proline uptake (system A; *c* and *d*) by phase-II human fibroblasts during in vitro ageing. Cells were grown for 22 h (non-confluent cultures; *a* and *c*) and for 7–14 days (confluent cultures; *b* and *d*). Transport assay was as described in figure 2. The data are expressed on a 'per cell' basis (■) or on a 'per milligram protein' basis (□). Each point is the mean of triplicate determinations. Lines are computer-derived best fits to a linear regression equation. Continuous lines = $p < 0.05$; broken lines = n.s.

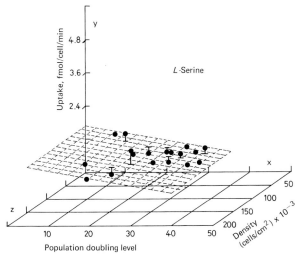

Fig. 4. L-Serine uptake (system ASC) by cultured human fibroblasts during in vitro ageing. Multiple regression surface of amino acid transport as dependent variable (y axis) versus ageing (population doubling level, x axis) and cell density (z axis) as independent variables. Data points are those given in figure 3a (non-confluent cultures) and figure 3b (confluent cultures), expressed as femtomoles per cell per minute. Bars are deviations of experimental points from theoretical points (on the plane) predicted by the multiple regression equation.

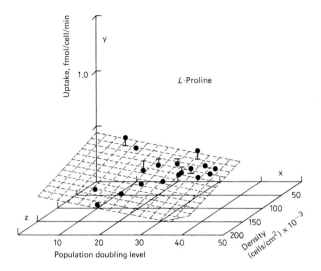

Fig. 5. L-Proline uptake (system A) by cultured human fibroblasts during in vitro ageing. Multiple regression surface calculated as in figure 4. Data points are those given in figure 3c (non-confluent cultures) and figure 3d (confluent cultures), expressed as femtomoles per cell per minute. For the meaning of the bars, see figure 4.

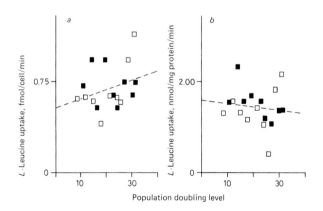

Fig. 6. L-Leucine uptake (system L) by cultured human fibroblasts during in vitro ageing. Cells were grown for 22 h (non-confluent cultures, □) or for 7–14 days (confluent cultures, ■). Transport assay was as described in figure 2. The data are expressed on a 'per cell' basis *(a)* or on a 'per milligram protein' basis *(b)*. Each point is the mean of triplicate determinations. Lines are computer-derived best fits to a linear regression equation. Both regressions are not significant.

$p < 0.01$), but its deviance was comparably contributed by ageing ($F = 8.9$; $p < 0.01$) and by cell density ($F = 9.7$; $p < 0.01$). This analysis indicates that both ageing and cell density affect amino acid transport activity by system A in cultured human fibroblasts. Similar conclusions were reached when the corresponding data expressed on a 'per milligram protein' basis were analyzed.

Transport activity of system L, as measured by *L*-leucine uptake, did not change significantly with ageing both in non-confluent and in confluent cultures. All the data were therefore plotted together and shown in figure 6 on a 'per cell' basis (fig. 6a) and on a 'per milligram protein' basis (fig. 6b). Multiple regression analysis applied to these data failed to reveal any significant change in transport activity by system L as a function of ageing or cell density (not shown).

Discussion

In multicellular organisms, the specialization of metabolic functions requires a continuous exchange of 'nutrients' among cells and tissues by way of the extracellular fluids and much of metabolism is coupled directly or indirectly to transport [*Elbrink and Bihler*, 1975; *Christensen*, 1982]. Alterations in transport activities may play a role in the performance of the cell cycle [*Cunningham and Pardee*, 1969; *Pardee* et al., 1974], in the control of cell growth and tissue development [*Holley*, 1975; *Christensen*, 1973] and in neoplastic progression [*Holley*, 1972; *Holley and Kiernan*, 1974]. As far as ageing is concerned, recent investigations [*Polgar* et al., 1978; *Germinario* et al., 1980; *Cremer* et al., 1981] indicate that the transport of sugars and nucleosides undergo no gross decrements on a 'per cell' basis when studied in cultured human fibroblasts. Conversely, the results recounted in this paper clearly show that the transport of neutral amino acids does indeed decrease with ageing either when expressed on a 'milligram protein' or on a 'per cell' basis. Not only phase-III fibroblasts exhibit a marked decrease in transport activity of the Na^+-dependent transport systems ASC and A, but a progressive decline of their activity is detectable as phase II proceeds. Phase III is characterized by an overall reduction in transport activity for all the systems studied (ASC, A and L). The fact that even the Na^+-independent system L is involved in this transport defect, whereas its activity does not change during phase II, suggests that a profound derangement of membrane permeability takes place in phase III. That this transport failure is not due to the presence of non-viable cells in the fibroblast population is indicated by experiments performed with phase-III cells after a short period following trypsinization (20 h) and attachment to the substratum, which yielded comparable results (not shown).

A decrease in the activity of the Na^+-dependent transport systems with ageing, here described for a cell population maintained under in vitro conditions, has also been observed using synaptosomes obtained from rats of increasing age [*Wheeler*, 1980, 1982]. It must be recognized, however, that the Na^+-dependent systems investigated in the latter studies were reactive towards different amino acids as glutamate (anionic) and Gaba (γ-amino acid).

An additional aspect of our study deserves consideration. The results presented in figures 3–5 show that the transport of *L*-serine (system ASC) is not substantially affected by cell density within the range encompassed by our culture conditions, whereas the transport of *L*-proline (system A) exhibits a definite dependence on this parameter. The apparent insensitivity of system ASC to cell density appears in contrast with the marked dependence of *L*-serine uptake by fibroblast-like cell lines [*Guidotti* et al., 1978; *Borghetti* et al., 1980; *Piedimonte* et al., 1982] and of *L*-glutamate uptake via system ASC by cultured human fibroblasts [*Dall'Asta* et al., 1983]. However, this difficulty can be overcome when one considers that, in our experiments, cell density varies within a narrow range and that this range lies in a region in which the density of the culture has only a little influence on transport activity by system ASC [*Piedimonte* et al., 1982]. This may not be true for system A, if its dependence on cell density were still rather strong in the same density region.

Further information on the biologic meaning of the changes in transport activity with ageing and on their underlying molecular mechanisms will be obtained in studies, now in progress, with fibroblast populations from donors of different age.

References

Bemiller, P.M.; Miller, J.E.: Cytological changes in senescing WI-38 cells: a statistical analysis. Mech. Age. Dev. *10:* 1–15 (1979).

Berumen, L.; Macieira-Coelho, A.: Changes in albumin uptake during the lifespan of human fibroblasts in vitro. Mech. Age. Dev. *6:* 165–172 (1977).

Borghetti, A.F.; Piedimonte, G.; Tramacere, M.; Severini, A.; Ghiringhelli, P.; Guidotti, G.G.:

Cell density and amino acid transport in 3T3, SV3T3 and SV3T3-revertant cells. J. cell. Physiol. *105:* 39–49 (1980).

Chen, J.J.; Brot, N.; Weissbach, H.: RNA and protein synthesis in cultured human fibroblasts from donors of various ages. Mech. Age. Dev. *13:* 285–295 (1980).

Christensen, H.N.: On the development of amino acid transport systems. Fed. Proc. *32:* 19–28 (1973).

Christensen, H.N.: Interorgan amino acid nutrition. Physiol. Rev. *62:* 1193–1233 (1982).

Cremer, T.; Werdan, K.; Stevenson, A.F.G.; Lehner, K.; Messerschmidt, O.: Aging in vitro and D-glucose uptake kinetics of diploid human fibroblasts. J. cell. Physiol. *106:* 99–108 (1981).

Cunningham, D.D.; Pardee, A.B.: Transport changes rapidly initiated by serum addition to 'contact inhibited' 3T3 cells. Proc. natn. Acad. Sci. USA *64:* 1049–1056 (1969).

Dall'Asta, V.; Gazzola, G.C.; Franchi-Gazzola, R.; Bussolati, O.; Longo, N; Guidotti, G.G.: Pathways of L-glutamic acid transport in cultured human fibroblasts. J. biol. Chem. *258:* 6371–6379 (1983).

Elbrink, J.; Bihler, I.: Membrane transport: its relation to cellular metabolic rates. Science *188:* 1177–1184 (1975).

Franchi-Gazzola, R.; Gazzola, G.C.; Dall'Asta, V.; Guidotti, G.G.: The transport of alanine, serine and cysteine in cultured human fibroblasts. J. biol. Chem. *257:* 9582–9587 (1982).

Gazzola, G.C.; Dall'Asta, V.; Franchi-Gazzola, R.; White, M.F.: The cluster-tray method for rapid measurement of solute fluxes in adherent cultured cells. Analyt. Biochem. *115:* 368–374 (1981a).

Gazzola, G.C.; Dall'Asta, V.; Guidotti, G.G.: The transport of neutral amino acids in cultured human fibroblasts. J. biol. Chem. *255:* 929–936 (1980).

Gazzola, G.C.; Dall'Asta, V.; Guidotti, G.G.: Adaptive regulation of amino acid transport in cultured human fibroblasts. J. biol. Chem. *256:* 3191–3198 (1981b).

Germinario, R.J.; Oliveira, M.; Taylor, M.: Studies on the effects of in vitro ageing on saturable and nonsaturable sugar uptake in cultured human skin fibroblasts. Gerontology *26:* 181–187 (1980).

Goldstein, S.; Stotland, D.; Cordeiro, R.A.J.: Decreased proteolysis and increased amino acid efflux in aging human fibroblasts. Mech. Age. Dev. *5:* 221–233 (1976).

Grinna, L.S.: Changes in cell membranes during aging. Gerontology *23:* 452–464 (1977).

Guidotti, G.G.; Borghetti, A.F.; Gazzola, G.C.: The regulation of amino acid transport in animal cells. Biochim. biophys. Acta *515:* 329–366 (1978).

Hayflick, L.: Cell aging. Annu. Rev. Gerontol. Geriat. *1:* 26–71 (1980).

Hayflick, L.; Moorhead, P.S.: The serial cultivation of human diploid cell strains. Expl Cell Res. *25:* 585–621 (1961).

Hill, B.T.: The establishment of criteria for 'quiescence' in ageing human embryo cell cultures and their response to a proliferative stimulus. Gerontology *23:* 245–255 (1977).

Holley, R.W.: A unifying hypothesis concerning the nature of malignant growth. Proc. natn. Acad. Sci. USA *69:* 2840–2841 (1972).

Holley, R.W.: Control of growth of mammalian cells in cell culture. Nature, Lond. *258:* 487–490 (1975).

Holley, R.W.; Kiernan, J.A.: Control of the initiation of DNA synthesis in 3T3 cells: serum factors. Proc. natn. Acad. Sci. USA *71:* 2908–2911 (1974).

Macieira-Coelho, A.; Pontén, J.; Philipson, L.: The division cycle and RNA-synthesis in diploid human cells at different passage levels in vitro. Expl Cell Res. *42:* 673–684 (1966).

Pardee, A.B.; Jiménez de Asua, L.; Rozengurt, E.: Functional membrane changes and cell growth: significance and mechanism; in Clarkson, Baserga, Control of proliferation in animal cells, pp. 547–561 (Cold Spring Harbor Laboratory, Cold Spring Harbor 1974).

Piedimonte, G.; Borghetti, A.F.; Guidotti, G.G.: Effect of cell density growth rate and amino acid transport in simian virus 40-transformed 3T3 cells. Cancer Res. *42:* 4690–4693 (1982).

Polgar, P.; Taylor, L.; Brown, L.: Plasma membrane associated metabolic parameters and the aging of human diploid fibroblasts. Mech. Age. Dev. *7:* 151–160 (1978).

Turk, B.; Milo, G.E.: An in vitro study of senescent events of human embryonic lung (WI-38) cells. Archs Biochem. Biophys. *161:* 46–53 (1974).

Wang, C.S.; Smith, R.L.: Lowry determination of protein in presence of Triton X-100. Analyt. Biochem. *63:* 414–417 (1975).

Wang, K.M.; Rose, N.R.; Bartholomew, E.A.; Balzer, M.; Berde, K.; Foldvary, M.: Changes of enzymatic activities in human diploid cell line WI-38 at various passages. Expl Cell Res. *61:* 357–364 (1970).

Wheeler, D.D.: Aging of membrane transport mechanisms in the central nervous system – high affinity glutamic acid transport in rat cortical synaptosomes. Exp. Gerontol. *15:* 269–284 (1980).

Wheeler, D.D.: Aging of membrane transport mechanisms in the central nervous system – 2-deoxy-D-glucose transport in cortical synaptosomes from the Long-Evans rat. Exp. Gerontol. *16:* 247–252 (1981).

Wheeler, D.D.: Aging of membrane transport mechanisms in the central nervous system – GABA transport in rat cortical synaptosomes. Exp. Gerontol. *17:* 71–85 (1982).

Zs.-Nagy, I.: The role of membrane structure and function in cellular aging: a review. Mech. Age. Dev. *9:* 237–246 (1979).

Dr. Gian C. Gazzola, Istituto di Patologia Generale, Università di Parma, Via Gramsci 14, I-43100 Parma (Italy)

Band 3, the Predominant Transmembrane Polypeptide, Undergoes Proteolytic Degradation as Cells Age[1]

Marguerite M.B. Kay

Division of Geriatric Medicine, Department of Medicine, and Departments of Biochemistry, and Microbiology and Immunology, Texas A&M University, Olin E. Teague Veterans' Center, Temple, Tex., USA

Introduction

Band 3, the major transmembrane polypeptide, mediates the exchange of anions (chloride and bicarbonate) across the membrane [1, 2] and appears to be the binding site for hemoglobin [3] and the glycolytic enzymes glyceraldehyde-3-P-dehydrogenase [4], aldolase [5], and phosphofructokinase [6]. Band 3 binds to band 2.1 which simultaneously binds to spectrin thereby linking the cytoplasmic surface of the plasma membrane to the internal filamentous cytoskeleton [7]. In addition, it may contain the glucose transport protein [8, 9], and is thought to be involved in the transfer of water across the membrane [10]. Band 3 has been demonstrated on all cells examined including fibroblasts, white blood cells, platelets, and liver cells [11, 12]. This multifunctional membrane protein appears to generate the senescent cell antigen which terminates the life of cells [13–23].

The senescent cell antigen is a $\sim 62,000$ M_r glycoprotein that appears on the surface of senescent and damaged cells [13–23]. It is recognized by the antigen binding Fab region [15, 16] of a specific immunoglobulin (Ig) G autoantibody in serum which attaches to the cells and initiates their removal by macrophages [14, 15]. The senescent cell antigen was first observed on the surface of senescent human erythrocytes [13, 14], but has since been demonstrated on the surface of lymphocytes, polymorphonuclear leukocytes, platelets, embryonic kidney cells, and adult liver cells [16, 21].

---

[1] This work was supported, in part, by the Veterans' Administration Research Service, NIH Grant No. AM32094 (subcontract No. 1058SC with UCSF), and DAMD 17-83-C-3165.

Since mature erythrocytes cannot synthesize proteins, the senescent cell antigen was thought to be generated by modification of a preexisting protein of higher molecular weight [18, 21]. It was postulated that the senescent cell antigen was a component of the band 4.5 region that was derived from band 3 [18, 21–23] based on both extraction and isolation conditions, relative molecular weight, and its characterization as a glycosylated peptide [16, 21].

Experiments designed to test this hypothesis revealed that the senescent cell antigen is immunologically related to band 3 and may represent a physiologically significant proteolytic product of the parent molecule [21–23]. Both band 3 and senescent cell antigen abolished the phagocytosis inducing ability of IgG eluted from senescent cells; whereas, spectrin, bands 2.1, 4.1, actin, glycohorin A, PAS staining bands 1–4, and desialylated glycophorin A and PAS staining bands 1–4 did not [19–21, 23]. In addition, monospecific antibodies to both purified band 3 and the senescent cell antigen reacted with band 3 and its proteolytic products as determined by immunoautoradiography of red blood cell (RBC) membranes indicating that these molecules share common antigenic determinants not possessed by other red cell membrane components [21, 23].

Since the senescent cell antigen shares antigenic determinants with band 3 and may be derived from it [21, 23], the effect of cellular age on the appearance of band 3 proteolytic products was investigated using both the gel overlay and immunoblotting techniques.

Materials and Methods

Cell Separation. RBC were separated into young, middle-aged, and old populations on Percoll (Pharmacia) gradients as previously described [17]. 4 ml of blood were mixed with 20 ml of Percoll diluted 1:10 with 10× Dulbecco's phosphate-buffered saline (PBS). Gradients were centrifuged at 18,450 g_{av} (25,950 g_{max}) for 30 min at 4 °C. Platelets, white cells, and reticulocytes formed bands at the top and were removed. Young RBC were in the least dense fraction (ρ 1.090), middle-aged cells in fractions ρ 1.10–1.11, and old cells in the most dense fraction (ρ 1.120) as determined by ^{59}Fe labeling in situ [17]. Old cells used for the studies described here represented 0.6% of the total cells. Middle-aged cells from the denser middle-aged fractions were used. Red cells were washed 3 times in 20 volumes of PBS by centrifugation at 3,000 g for 10 min, lyzed, and the membranes washed with 5 mM sodium phosphate buffer, pH 7.4, containing 1 mM ethylenediaminetetraacetic acid (EDTA), 1 mM ethyleneglycol-bis-(β-aminoethylether)N,N'-tetraacetic acid (EGTA), 5 mM diisopropylfluorophosphate (DFP) and 100 µg/ml phenylmethylsulfonyl fluoride (PMSF) as protease inhibitors.

Isolation of IgG from Senescent RBC. IgG was isolated as previously described [15, 16]. Briefly, middle-aged and old RBC were washed 3× with 50–100 volumes of PBS, pH 7.4. RBC membranes were prepared by digitonin lysis, washed 3× with PBS, and IgG eluted with 0.1 M

glycine-HCl buffer, pH 2.3. Eluates were neutralized with 1 N NaOH and concentrated using an Amicon Diaflow with a PM 10 filter. IgG was isolated from eluates by affinity chromatography with protein A Sepharose 4B [15, 16, 18].

^{125}I Labeling of RBCs. Middle-aged red cells were washed 5× with 20 volumes of PBS and resuspended in Alsevers' solution. They were incubated overnight at 4°C on a rotator, centrifuged, and resuspended to a hematocrit of ~70% in Alsevers' solution. ^{125}I Bolton-Hunter reagent (2.3 mCi; New England Nuclear) was added to 5 cm³ of the cell suspension at ice temperature. After 2 h, 45 cm³ of Hank's buffered salt solution were added, and the cells incubated at ice temperature overnight. Cells were washed 4× with PBS and membranes prepared as described earlier.

Rabbit Antisera to Band 3. Antibodies to band 3 were prepared by injecting rabbits one a month with purified band 3 [23]. Antisera to band 3 was absorbed with peripheral membrane proteins obtained by treating ghosts with 0.1 N NaOH, and with PAS staining bands 1–4 as previously described [23]. Proteins and glycoproteins used for absorption were covalently coupled to Sepharose 4B [23]. This was done to ensure specificity although binding to peripheral membrane proteins or PAS staining bands 1–4 was not observed prior to absorption [23].

Sodium Dodecyl Sulfate (NaDodSO$_4$) Polyacrylamide Gel Electrophoresis. Proteins were analyzed on two different gel systems: 4–30% polyacrylamide gradient gels containing NaDodSO$_4$ and utilizing a continuous buffer system [16, 21, 23], and 7% polyacrylamide gels containing NaDodSO$_4$ using the discontinuous buffer system of *Laemmli* [24]. A more complete description of the electrophoresis conditions has previously appeared [16, 23]. The sample buffer was 10 mM Tris-HCl, pH 6.8, containing 1 mM EDTA, 40 mM dithiothreitol (DTT), 5% glycerol, 2% NaDodSO$_4$, and 0.2 mg/ml pyronin Y. Gels were stained for protein with Coomassie blue and for glycoproteins with dansyl hydrazine.

Immunostaining of Membrane Proteins. Immunoautoradiography was performed by the gel overlay method of *Granger and Lazarides* [25] with minor modifications [23] or by the immunoblotting technique [26] with the modifications described previously [23]. Briefly, membrane proteins were separated on 4–30% polyacrylamide gradient gels, and transferred electrophoretically to nitrocellulose sheets [26]. Free binding sites on the nitrocellulose sheets were blocked by incubation with 3% bovine serum albumin in 0.9% NaCl (1 h, 24°C). The paper was incubated with antisera to membrane proteins or IgG eluted from senescent cells. After washing with 0.9% NaCl containing 0.05% Tween 20, blots were incubated with ^{125}I-labeled protein A (New England Nuclear, 70–100 μCi/μg) in 3% bovine serum albumin for 2 h at 24°C. 'Immunoblots' were washed, dried, and exposed to Kodak X-Omat RP film at −80°C for 1–10 days in a Kodak cassette with intensifying screen. Transfer of polypeptides was monitored by loss of Coomassie blue-staining bands from the gel, and by the appearance of amido black staining bands on the nitrocellulose paper. Transfer of polypeptides was >90% efficient. Neither preimmune serum nor protein A bind to red cell proteins under the conditions employed [23].

Results

IgG Eluted from Senescent RBC Binds to Band 3 and Its ~62,000 Dalton Degradation Product. In order to determine whether IgG eluted from senescent cells recognized a specific degradation product of band 3, erythrocyte

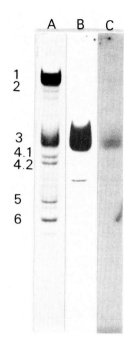
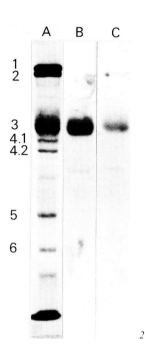

Fig. 1. Binding of antisera to band 3 and IgG eluted from senescent cells to band 3 and its proteolytic breakdown product. Lanes: A = Coomassie blue stain of RBC membrane proteins; B = immunoautoradiography with anti-band 3; C = immunoautoradiography with IgG eluted from senescent red cells. Old red cell membrane proteins were separated on sodium dodecyl sulfate-polyacrylamide gradient gels. Polypeptides were visualized by Coomassie blue staining. Polypeptides to which rabbit antibodies to band 3 or human IgG eluted from senescent RBC bind were identified by electrophoretic transfer from polyacrylamide gels to nitrocellulose paper, followed by overlay of the nitrocellulose paper with antisera and then ^{125}I-labeled protein A.

Fig. 2. ^{125}I labeling of the ~62,000 dalton proteolytic product of band 3 in intact cells. Lanes: A = Coomassie blue stain of RBC membrane proteins; B = autoradiography of membrane proteins from red cells labeled with ^{125}I Bolton-Hunter reagent; C = immunoautoradiography with anti-band 3. Middle-aged red cells were washed and incubated with ^{125}I Bolton-Hunter reagent as described in the text. Membranes were prepared and analyzed by polyacrylamide gel electrophoresis (lane B). The gel was dried and exposed to film. Membranes from cells that were not labeled were run at the same time on the same slab gel and incubated with antibodies to band 3 followed by ^{125}I protein A as described in the text and the legend to figure 3 (lane C). Autoradiographs shown in lanes B and C were both exposed to film for 5 days.

membrane proteins were transferred from $NaDodSO_4$/4–30% polyacrylamide gels to nitrocellulose paper. The paper was overlaid with IgG eluted from senescent cells followed by incubation with ^{125}I-labeled protein A. Autoradiographs obtained by exposing the nitrocellulose paper to X-ray film revealed binding of senescent cell IgG to a polypeptide migrating at M_r ~ 62,000 (± 3%; range 60,000–64,000) which corresponded to the same M_r ~ 62,000 polypeptide labeled with monospecific antibodies to band 3 (fig. 1). These results suggest that the antigenic determinants recognized by the IgG eluted from senescent cells reside on the M_r ~ 62,000 fragment of band 3.

Band 3 Degradation Product M_r ~ 62,000 Communicates with the Extracellular Space. As an approach to determining whether the M_r ~ 62,000 polypeptide of band 3 is exposed on the outside of whole cells, intact red cells were labeled with ^{125}I Bolton-Hunter reagent. Membranes were prepared, and analyzed by polyacrylamide gel electrophoresis. Autoradiography of Coomassie blue stained gels revealed labeling of band 3, and the polypeptide migrating at M_r ~ 62,000 (fig. 2). In addition, both band 3 and the M_r ~ 62,000 polypeptide labeled with antibodies to band 3 (fig. 2). These findings indicate that the ~ 62,000 dalton fragment of band 3 has a segment that is exposed to the extracellular space.

Band 3 Degradation Products Increase with Cell Age. Young, middle-aged, and old cells were separated on Percoll gradients. Membranes were prepared and erythrocyte proteins were separated by polyacrylamide gel electrophoresis. Antibodies to band 3 were used to determine the relative amount of band 3 breakdown product in the membranes of young, middle-aged, and old cells by immunoautoradiographic gel staining [11]. Binding of antibodies to band 3 and its ~ 62,000 M_r breakdown product was detected in middle-aged and old cell membranes by autoradiography (fig. 3). In contrast, the ~ 62,000 M_r polypeptide of band 3 was not detected in autoradiographs of membranes from young cells although band 3 was labeled (fig. 3). Thus, the ~ 62,000 M_r breakdown product of band 3 is detected in middle-aged and old cells, but not in young cells.

Discussion

IgG eluted from senescent red cells reacts with band 3 and its M_r ~ 62,000 breakdown product. This breakdown product is present in red cell

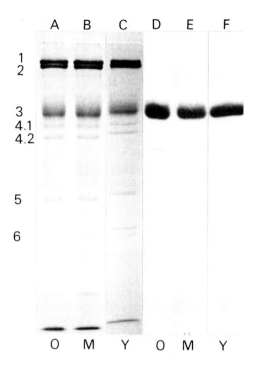

Fig. 3. Detection of band 3 proteolytic breakdown products in red cell membranes prepared from middle-aged and old cells. Lanes: A–C = Coomassie blue stain of RBC membrane proteins; D–F = immunoautoradiographs; A, D = old; B, E = middle-aged; C, F = young. RBC were separated into young, middle-aged, and old populations on the basis of density, washed, and lyzed. Processing time from the time blood was drawn until membranes were in electrophoresis sample buffer was < 3 h. Membrane proteins were separated on NaDodSO$_4$ polyacrylamide slab gel using the discontinuous system of *Laemmli* [24]. The gel was fixed in 50% ethanol, 10% acetic acid (6 h, 22°C), and then neutralized in three changes of 50 mM Tris-HCl, 100 mM NaCl, 5 mM NaN$_3$, pH 7.5 for 24 h. The gel was washed in buffer 1 (10 mM Tris-Cl, 140 mM NaCl, 5 mM NaN$_3$, 0.1 mM EDTA, 0.1% gelatin (pH 7.5) for 24 h, followed by a 24-hour incubation in buffer 1 to which anti-band 3 was added (1:1,000 dilution with buffer 1). Unbound antibody was removed by 3 changes of buffer 1 over 3 days at 22°C. The gel was incubated in buffer 1 containing 2.5 µCi of ^{125}I-protein A (New England Nuclear) for 24 h at 22°C. Unbound ^{125}I-protein A was removed by 2 days of washing with buffer 1 and one day of washing with buffer 1 without gelatin. The gel was stained with 0.1% Coomassie blue, 47.5% ethanol, 10% acetic acid, and destained in 12.5% ethanol, 5% acetic acid. The gel was dried, and autoradiographed for 5 days (−70°C, with intensifying screen).

membranes prepared with EDTA, EGTA, DFP, and PMSF to avoid antifactual proteolysis as described here and previously [23]. The $\sim 62,000$ dalton band 3 degradation product recognized by IgG eluted from senescent cells is not detected in membranes prepared from young cells but is present in the membranes of old cells.

Since the $\sim 62,000$ M_r polypeptide labels with ^{125}I when intact cells are incubated with Bolton-Hunter reagent, the $\sim 62,000$ M_r breakdown product appears to contain a segment that is exposed to the extracellular space. Trace amounts of a $\sim 60,000$ dalton degradation product of band 3 have been observed by others in purified band 3 preparations even when several different isolation procedures were used [27–29]. *Tarone* et al. [29] observed that breakdown products of band 3 increased when washed membranes were stored for several days at 4°C without protease inhibitors. Storage induced degradation of band 3 was reported to be stepwise with production of 40,000 and 24,000 dalton peptides followed by breakdown into smaller peptides of $\sim 30,000$ and 12,000 M_r [29], and was independent of leukocyte and platelet contamination [29].

The $\sim 62,000$ dalton breakdown product of band 3 described here is present in old but not young red cell membranes prepared with the protease inhibitors DFP, PMSF, EDTA and EGTA. Therefore, it appears that band 3 undergoes proteolysis in vivo as cells age. Damage to the multifunctional membrane protein band 3 might explain many observations of cellular aging including age-related cellular dehydration, loss of cell deformability, and increased fragility [30–33].

References

1. Cabantchik, Z.I.; Rothstein, A.: Membrane proteins related to anion permeability of human red blood cells. I. Localization of disulfonic stilbene binding sites in proteins involved in permeation. J. Membr. Biol. *15:* 207–226 (1974).
2. Lepke, S.; Fasold, H.; Pring, M.; Passow, H.J.: A study of the relationship between inhibition of anion exchange and binding to the red blood cell membrane of 4,4′-diisothiocyano-2,2′-disulfonic acid (DIDS) and its dihydro-derivative (H_2DIDS). J. Membr. Biol. *29:* 147 (1976).
3. Salhany, J.M.; Shaklai, N.: Functional properties of human hemoglobin bound to the erythrocyte membrane. Biochemistry *18:* 893–899 (1979).
4. Kliman, H.J.; Steck, T.L.: Association of glyceraldehyde-phosphate dehydrogenase with the human red cell membrane a kinetic analysis. J. biol. Chem. *255:* 6314–6321 (1975).
5. Strapazon, E.; Steck, T.L.: Interaction of the adolase and membrane of human erythrocytes. Biochemistry *16:* 2966–2971 (1977).

6   Karadsheh, N.S.; Uyeda, K.; Oliver, R.M.: Studies on structure of human erythrocyte phosphofructokinase. J. biol. Chem. *252:* 3515–3524 (1977).

7   Bennett, V.; Stenbuck, P.J.: The membrane attachment protein for spectrin is associated with band 3 in human erythrocyte membrane. Nature, Lond. *280:* 468–473 (1979).

8   Taverna, R.D.; Langdon, R.G.: *D*-Glucosyl isothiocyanate, an affinity label for the glucose transport proteins of the human erythrocyte membrane. Biochem. biophys. Res. Commun. *54:* 593–599 (1973).

9   Lin, S.; Spudich, J.A.: Binding of cytochalasin B to red cell membrane protein. Biochem. biophys. Res. Commun. *61:* 1471–1474 (1974).

10 Brown, P.A.; Feinstein, M.B.; Shuaki, R.I.: Membrane proteins related to water transport in human erythrocytes. Nature, Lond. *254:* 523–525 (1975).

11 Kay, M.M.B.; Goodman, S.R.; Cone, C.: A polypeptide immunologically related to red cell band 3 is present in white cells. Blood *60:* 64a (1982).

12 Kay, M.M.B.; Tracey, C.; Goodman, J.; Bassel, P.; Cone, C.: Polypeptides immunologically related to erythrocyte band 3 are present in nucleated somatic cells. Proc. natn. Acad. Sci. USA (in press, 1983).

13 Kay, M.M.B.: Mechanism of macrophage recognition of senescent red cells. Gerontologist *14:* 33 (1974).

14 Kay, M.M.B.: Mechanism of removal of senescent cells by human macrophages in situ. Proc. natn. Acad. Sci. USA *72:* 3521–3525 (1975).

15 Kay, M.M.B.: Role of physiologic autoantibody in the removal of senescent human red cells. J. supramol. Struct. *9:* 555–567 (1978).

16 Kay, M.M.B.: Isolation of the phagocytosis-inducing IgG-binding antigen on senescent somatic cells. Nature, Lond. *289:* 491–494 (1981).

17 Bennett, G.D.; Kay, M.M.B.: Homeostatic removal of senescent murine erythrocytes by splenic macrophages. Exp. Hematol., Copenh. *9:* 297–307 (1981).

18 Kay, M.M.B.: Molecular aging: a termination antigen appears on senescent cells; in Peeters, XXIX Annu. Coll. Protides of the Biological Fluids, Brussels 1981, vol. 29, pp. 325–328 (Pergamon Press, Oxford 1982).

19 Kay, M.M.B.: The senescent cell antigen is not a disialylated glycoprotein. Blood *58:* 90a (1981).

20 Kay, M.M.B.; Bennett, G.D.: Letter to the editor. Blood *59:* 1111–1112 (1982).

21 Kay, M.M.B.; Sorensen, K.; Wong, P.; Bolton, P.: Antigenicity, storage and aging: physiologic autoantibodies to cell membrane and serum proteins. Mol. cell. Biochem. *49:* 65–85 (1982).

22 Kay, M.M.B.; Goodman, S.; Sorensen, K.; Whitfield, C.; Wong, P.; Zaki, L.; Rudloff, V.: The senescent cell antigen is immunologically related to band 3. J. Cell Biol. *95:* 244a (1982).

23 Kay, M.M.B.; Goodman, S.; Sorensen, K.; Whitfield, C.; Wong, P.; Zaki, L.; Rudloff, V.: The senescent cell antigen is immunologically related to band 3. Proc. natn. Acad. Sci. USA *80:* 1631–1635 (1983).

24 Laemmli, U.K.: Cleavage of structural proteins during the assembly of the head of bacteriophage T4. Nature, Lond. *227:* 680–685 (1970).

25 Granger, B.L.; Lazarides, E.: Synemin: a new high molecular weight protein associated with desmin and vimentin filaments in muscle. Cell *22:* 727–738 (1980).

26 Towbin, H.; Staehelin, T.; Gordon, J.: Electrophoretic transfer of proteins from polyacryl-

amide gels to nitrocellulose sheets: procedure and some applications. Proc. natn. Acad. Sci. USA 76: 4350–4354 (1979).
27 Yu, J.; Steck, S.L.: Isolation and characterization of band 3, the predominant polypeptide of the human erythrocyte membrane. J. biol. Chem. 250: 9170–9175 (1975).
28 England, B.J.; Gunn, R.B.; Steck, T.L.: An immunological study of band 3, the anion transport protein of the human red blood cell membrane. Biochim. biophys. Acta 623: 171–182 (1980).
29 Tarone, G.; Hamaski, N.; Fukuda, M.; Marchesi, V.T.: Proteolytic degradation of human erythrocyte band 3 by membrane associated protease activity. J. Membr. Biol. 48: 1–12 (1979).
30 Piomelli, S.; Lurinsky, G.; Wasserman, L.R.: The mechanism of red cell aging. 1. Relationship between cell age and specific gravity evaluated by ultra-centrifugation in a discontinuous density gradient. J. Lab. clin. Med. 69: 659–674 (1967).
31 Cohen, N.S.; Ekholm, J.E.; Luthra, M.G.; Hanahan, D.J.: Biochemical characterization of density-separated human erythrocytes. Biochim. biophys. Acta 419: 229–242 (1976).
32 Williams, A.R.; Morris, D.R.: The internal viscosity of the human erythrocyte may determine its lifespan in vivo. Scand. J. Haematol. 24: 57–62 (1980).
33 Detraglia, M.; Cook, F.B.; Stasiw, D.M.; Cerny, L.L.: Erythrocyte fragility in aging. Biochim. biophys. Acta 345: 213–219 (1974).

M.M.B. Kay, MD, Division of Geriatric Medicine, Department of Medicine,
Olin E. Teague Veterans Center, 1901 South First Street, Temple, TX 76501 (USA)

Mitochondrial DNA in *Podospora anserina*

A Molecular Approach to Cellular Senescence

Donald J. Cummings

Department of Microbiology and Immunology, University of Colorado Health Sciences Center, Denver, Colo., USA

Introduction

Cellular senescence in the ascomycete fungus *Podospora anserina* has been known for quite some time [1–3]. In many respects, the theories put forth by *Marcou* and *Rizet* were forerunners of those developed by *Holliday* et al. [4] for mammalian cell ageing. There are three primary features of the Rizet-Marcou analysis. First, cells exist in either a nonsenescent or senescent state and that a random event occurs resulting in 'commitment' to senescence. In other words, there is a defined program. Second, each race within a species can have characteristic programs or parameters of senescence. Third, that senescence is maternally inherited and may be transmitted by a cytoplasmic factor [5]. This last point is critical and was further developed by genetic and cytoplasmic inhibitor studies [6–8]. No other system so firmly provided a testable model for understanding cellular ageing.

In spite of these important considerations, no attempt was made to study the one known organelle common to most eukaryotic cells: the mitochondrion. In 1977, I elected to take Sabbatical leave at the Centre de Génétique Moléculaire, CNRS, in Gif-sur-Yvette, France, where *Léon Belcour* was amassing phenomenological evidence supporting the maternal inheritance concepts fostered by *Rizet* and *Marcou*. Our primary goal was to isolate and characterize mitochondrial DNA from nonsenescent and senescent cultures of *P. anserina*. We were able to show that during senescence, the 94 kilobasepair (kbp) circular mitochondrial genome was replaced by a multimeric set of small circular molecules [9, 10]. Subsequent work in my laboratory [11], and by others [12, 13] showed that these molecules consisted of amplified regions of the nonsenescent mitochondrial genome. Not only were these studies critical for conceptualizing cellular senescence in *P. anserina*, they ushered in molecular studies on other aspects of cytoplasmic abnormalities in *Neurospora crassa* [14] and *Aspergillus nidulans* [15]. My purpose here is to present molecular evidence which suggests that these senescent multimeric sets of mitochondrial DNA act as mobile elements which play an active role in determining senescence.

Materials and Methods

P. anserina Races and Preparation of Mitochondrial DNA

Two races were chosen for study: race A which undergoes the most rapid senescence (ca. 10 cm growth) and race s with an intermediate senescence program (ca. 30 cm) [3]. Restriction enzyme analysis showed that these two races differed by only one or two sites [9]. Mycelia were propagated at 27°C on solid agar plates supplemented with corn meal extract [8]. Liquid bulk culturing was done in corn meal extract containing 5 g/l Difco yeast extract [9]. Senescent liquid cultures were prepared utilizing as inocula radial sections cut 2–3 mm behind the growth stoppage front of senescent thalli [10]. Mitochondria were prepared as previously described [9, 10] using β-glucuronidase to weaken hyphal walls and a Tekmar tissue mizer to disrupt these weakened mycelia [16]. Mitochondrial (and nuclear) DNA was prepared using DAPI-CsCl as before [9, 10]. In each case, two cycles of gradient equilibration were employed.

DNA Hybridization

Southern analysis of homologous DNA fragments were conducted according to published procedures [16, 17].

DNA Sequencing

Maxam-Gilbert sequencing methods were used throughout [18]. DNA fragments were first dephosphorylated and then labeled at their 5' ends using polynucleotide kinase and ^{32}P-γATP(ICN). Uniquely labeled 5' ends were obtained by cleavage with an appropriate second enzyme.

Results

As indicated, our early results showed that for both races s and A, mitochondrial DNA from nonsenescent mycelia consisted of a 94 kbp circular molecule [9]. In the senescent state, a set of multimeric circular molecules was observed with a monomeric unit size of 2.6 kbp [10]. These results led to several immediate considerations. First, what was the origin of these molecules? Were they derived from the nuclear genome, mitochondrial or neither? Second, would other independent senescent events give rise to this same set of molecules or would different monomeric sets occur? Third, how could

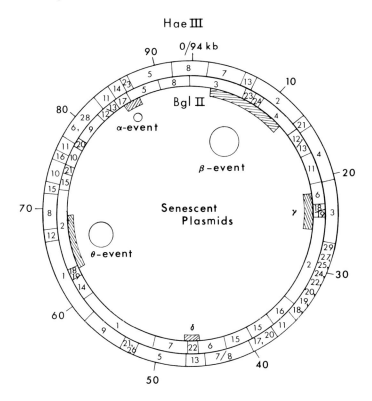

Fig. 1. Restriction maps of *Podospora anserina* mitochondrial DNA and the localization of senDNA. The outer circle represents the HaeIII endonuclease map and the inner circle that constructed from BglII endonuclease fragments. The three senescent mitochondrial DNA plasmids which have been cloned in their entirety (α, β and θ) are shown as circles of relative size (2.6, 9.8 and 6.3 kbp, respectively) [12, 21].

these molecules be involved with cellular senescence? With regard to this last point, *Belcour and Begel* [19] drew an analogy with suppressiveness in the petite mutation in yeast [20] where excision-amplification of specific regions of the mitochondrial genome 'suppressed' replication of the complete mitochondrial genome. To address these considerations, additional experiments were necessary, particularly with respect to the molecular characterization of *P. anserina* mitochondrial DNA.

Table I. Localization of P. anserina mitochondrial DNA structural genes

Yeast gene	Podospora probe[1]	Homologous restriction fragments	
		Hae III	Bgl II
21 SrDNA		3, several	2
15 SrRNA		2, 4, 21	4, 12/13
	21 SrRNA	Several	2
	15 SrRNA	2, 4, 21	4, 12/13
Oxi 1		5	7
Oxi 2		2	4
Oxi 3		14	17
Oli 1		ND	1
Oli 2		ND	2
Cobb		16	10
	α-senDNA	14, 23	5, 17
	β-senDNA	2, 7, 13	3, 4, 23, 24
	γ-senDNA	3	2, 6, 18/19

ND = Not determined.

[1] To verify the validity of the heterologous hybridization of the yeast mitochondrial DNA probes, rRNA from P. anserina was also utilized. The results obtained using several senDNAs isolated from independent senescent events are also presented.

Restriction Enzyme Mapping and Homology of Senescent DNA

Ordering the several DNA fragments from a 94 kbp genome was no simple task. This was greatly facilitated by the construction of a complete EcoRI clone bank [16]. Highly purified probes could then be prepared and applied to partial digestion products, other enzyme analyses, etc. Maps for six different enzymes were constructed and two of these (HaeIII and BglII) are displayed in figure 1. Superimposed on these maps are the results obtained for senescent DNA from several independent senescent events. From data gathered in several laboratories [10–13], there appears to be at least five unique sets of senescent DNA. The most prevalent of these, the first discovered, is termed α-senDNA. This DNA, as we will see later, hybridizes strongly to the HaeIII 14,23 region of the nonsenescent mitochondrial DNA and not at all to the nonsenescent nuclear genome. Other senescent events yielded different sets of senescent DNA (β, γ, δ, θ, etc.) and their localizations are also illustrated on the restriction maps. Clearly, senescent DNA can arise from several regions of the mitochondrial genome.

Two immediate questions arose from these findings. One, was the origin of these senDNAs truly random? And two, what was the nature of these molecules? Were they derived and replicated as a result of simple degradation of the mitochondrial DNA or were they excised from precise nucleotide sequences and then amplified? To answer the first of these questions, we [21] analyzed transcription products of the nonsenescent mitochondrial DNA. In addition, a genetic map was constructed using heterologous gene probes from *S. cerevisiae* [22]. We were able to show that for each senescent DNA, specific gene regions were involved: α-oxi 3; β-oxi 2; θ- an unidentified transcript; δ-oxi 1; and γ- the large ribosomal DNA (table I). Hence, even though the distribution of origin of these senescent DNAs appears to be random, specific nucleotide sequences unique to genes may be required. To obtain an answer to the second question, initially two approaches were used. First, several senescent events which yielded the same multimeric sets of DNA were analyzed. In each case the restriction enzyme fragment sizes were quite specific with no measurable differences in mobility. Second, and this was not always possible, the entire monomeric unit was cloned. This was successful for three unique senDNAs: α and β by us [21], and α and θ by *Belcour* et al. [23]. For each of these clones, the restriction maps were circular and were colinear with the nonsenescent mitochondrial DNA from which they were derived [21, 23]. Later, we will see that the nucleotide sequences themselves are identical for α-senDNA from races s and A. These results are critical since now we can state unequivocally that senDNA is excised and amplified by a specific mechanism and is not the result of some nonspecific degradative process.

Nuclear/Mitochondrial Interaction:
A Possible Role of senDNA in Senescence

The occurrence of these senDNAs in senescent mitochondria was so striking that further questions immediately came to mind. Most important, were these molecules the cause or effect of cellular ageing and by what mechanism? Two facts were apparent: these DNAs were of mitochondrial origin and they were autonomously replicated [24]. Could they be the transmissible cytoplasmic agents alluded to earlier [5]? To get at these questions two approaches were taken. For both, we utilized the most prevalent of these DNAs: α-senDNA.

First, we compared the time of appearance of α in races A and s [25]. In figure 2 it can be seen that even in nonsenescent mycelia, α is detectable not just at the HaeIII 23,14 position but as the free plasmid. For the more rapidly

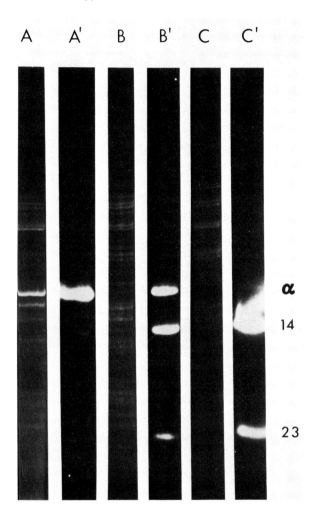

Fig. 2. Correlation of appearance of α-senDNA with program of senescence. Mitochondrial DNA (density 1.694 g/cm^3 [9]) was prepared and purified from senescent race A (lane A), nonsenescent race A (lane B) and nonsenescent race s (lane C). The DNA was digested with HaeIII restriction enzyme, examined on agarose gels [16] and hybridized to clone purified α-senDNA [16, 17]. As can be noted, ^{32}P-α-DNA hybridized to the major DNA fragment, isolated from senescent race A (lane A′). This same fragment can be observed in nonsenescent race A where α and HaeIII fragments 14 and 23 show essentially equimolar quantities (lane B′). In contrast, nonsenescent mitochondrial DNA from race s showed hybridization primarily to HaeIII fragments 14 and 23 and much less to α (ca. 20-fold, lane C′).

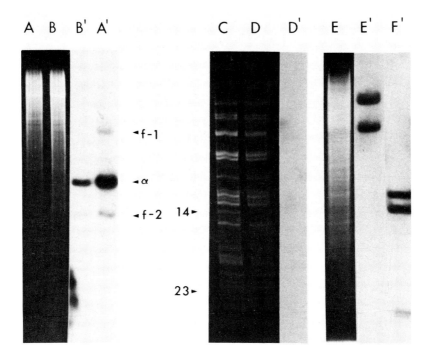

Fig. 3. Demonstration of the transfer of α-senDNA to the nuclear genome during senescence. Nuclear DNA (density 1.712 g/cm^3 [9]) was isolated as described. Lane A is s(+) senescent DNA cleaved with BglII enzyme and that in lane B s(−) senescent DNA also cleaved with BglII enzyme. There is one BglII site in α-senDNA [21]. Hybridization with ^{32}Pα-senDNA showed strong hybridization at the position of the linearized α-plasmid (lanes B' and A') as well as at two other positions, one of higher molecular weight (f-1) and the other lower (f-2) than the linearized α-plasmid. Lanes C and D show mitochondrial DNA isolated from race s and mex-1 digested with HaeIII restriction endonuclease. Note the absence of HaeIII fragments 14 and 23 in mex-1 mitochondrial DNA. That this does not imply simply the loss of a HaeIII restriction site is illustrated in lane D' where hybridization to ^{32}P-α-DNA was not detected. Lane E contains mex-1 nuclear DNA digested with Sal I endonuclease, another enzyme for which there is only one site in α-DNA [21]. Lane E' represents hybridization against ^{32}P-α-DNA where two high molecular weight fragments are detected (11 and 7 kbp, respectively). As expected, two fragments of lower molecular weight (ca. 3.5 and 2.5 kbp) were detected by hybridization against mex-1 nuclear DNA digested with HaeIII enzyme (lane F').

senescing race A, α is present at essentially equimolar concentrations as the mitochondrial genomic DNA, a much higher concentration than is seen in the less rapidly senescing race s. This suggests that there is a correlation between the onset of senescence and the appearance of senDNA.

Second, we determined whether α could function as a mobile element [25]. Regardless of whether senDNA actively 'suppressed' mitochondrial DNA replication or simply out-produced it, mitochondrial DNA is replaced during senescence and hence is not available for such a mobility study. But there is ample nuclear DNA. Senescent nuclear DNA was examined for the presence of α senDNA (and β) by hybridization techniques [17]. Using restriction endonuclease enzymes which cleaved or do not cleave α we were able to show α senDNA could be integrated into the senescent nuclear DNA (fig. 3). Senescent nuclear DNA obtained from either race s(+), s(−) or race A (not shown here) displayed hybridization to ^{32}P-labelled purified α-senDNA. For both, α appeared to be integrated in multiple copy with nuclear flanking sequences readily observed (f-1 and f-2, lane A'). It cannot be overemphasized that α is an integral part of the nonsenescent mitochondrial genome and is not detected in nonsenescent nuclear DNA. Other experiments showed that β could also be found in senescent nuclear DNA [25].

These results on the nuclear transfer of mitochondrial senDNA were considered so critical for understanding senescence that we had to be certain that they were not artifactual. Previously, *Vierny* et al. [26] had isolated some very interesting mutants of race s. These were isolated as outgrowths of senescent mycelia and were termed mex. One of these, mex-1, was examined in detail and it was shown that the HaeIII 23,14 region of the mitochondrial genome, from which α is excised, was absent. We obtained this mutant from *Belcour* and examined its nuclear DNA. As can be seen in figure 3, lane D, mex-1 mitochondrial DNA lacked HaeIII fragments 14 and 23 and hybridization to ^{32}Pα-DNA was not detected (D'). On the other hand, mex-1 nuclear DNA showed clear hybridization to α-DNA (lanes E' and F'). Mex-1 nuclear DNA was cleaved with either SalI or HaeIII restriction endonucleases. As one might expect from the predicted size of fragments generated by these different enzymes, SalI digested mex-1 nuclear DNA yielded homologous fragments of higher molecular weight than those from HaeIII digestion. Both enzyme digests yielded two homologous fragments, suggesting that α was integrated as single copy in mex-1 nuclear DNA. *Vierny* et al. [26] also showed that these mex mutants which lacked α in their mitochondrial genome experienced senescence at a negligible rate.

```
                                    J1
                                    ▼
                         GTTATATAAC
Hae23            5'   GGCCAAGTGTTCAATATATTGCAG

α-DNA    ←  CTATATAGACTAAGGACTGGCTGCTTATCCTAC  GTGCGCCGTTTAACGTGCGTTTTAAGTCCGG    3'  Msp
                                               TAATATATTA
                              ATAACCAATTATATAATAGCATCATTCAG    →    3'  Hae14
                                    ▲
                                    J2
```

Fig. 4. DNA sequence of the excision sites of α-plasmid from either race A or s were determined. Few differences were noted in DNA sequences between races A and s. The sequence for HaeIII 23 was identical with that of α until 24 bp from the left HaeIII site and then the sequence diverged (J1). Similarly, α and HaeIII fragment 14 sequences were identical until 172 bp from the right HaeIII site before divergence occurred (J2). Those divergent sequences left of J1 and right of J2 are shown as is the contiguous α-DNA sequence through the J1–J2 junction splice position. Note the 10 bp palindromic sequences on either side of the excision sites. Note also the 7 bp direct repeat at the splice junction.

DNA Sequence of α Excision Sites

Our contention is that senDNA is acting as a mobile element and its integration can have developmental ramifications. Since we had constructed clones of the non-senescent mitochondrial DNA and clones of α for both races A and s, we were in a position to determine the excision sites by direct comparison with homologous sequences as well as to decide whether these senDNA sequences were conserved [27].

In figure 4, the sequence of α at these excision sites is presented. The sequences of the contiguous mitochondrial DNA at both termini are also illustrated. There are several important features of these excision sequences. First, bordering the excision sites (J1 and J2) on the mitochondrial genomic DNA, there are 10 bp palindromic sequences. Each is distinct but structural similarities are apparent. These palindromic sequences could serve as recognition sites for an enzyme complex which holds together the two distal regions prior to excision and ligation. Second, at the junction site itself, there is a 5-bp sequence directly repeated 8 bp downstream, which is converted to a 7-bp repeated sequence (ACGTGCG) by the splicing of the junction sites. This sequence could be the target site for ligation. Third, in the entire sequence of α, some 2,600 bp, only three base changes have been found between races s and A. Thus, α-senDNA is highly conserved.

Discussion

Like most eukaryotes, *P. anserina* undergoes cellular ageing. We have presented here a review of our findings for a molecular approach to understanding the senescent process. Specific gene regions of the mitochondrial genome are excised, amplified and are isolatable as freely replicating plasmids. In a rapidly senescing cell these so-called senDNA plasmids are detected at an earlier stage than in less rapidly senescing cells. Examination of senescent nuclear DNA (but *not* nonsenescent) showed that these mitochondrially derived plasmids could integrate into the nuclear genome. Finally, DNA sequence analysis revealed that the excision sites are highly conserved and that short repeated sequences exist which could serve as either recognition or target sites for excision.

These findings are quite provocative. They suggest that in *P. anserina*, senescence may involve the transmission of mobile elements which are integral parts of mitochondrial gene regions. This interpretation has several parallels in other developmental abnormalities. *Pall* [28], for example, has shown that extrachromosomally replicating plasmid-like molecules are detected in certain cancers. More specifically, in crown gall disease, neoplasia is a direct result of the insertion of a segment of the Agrobacterium Ti plasmid [29]. In attempting to determine the mechanism for the mobility of mitochondrial elements described here, three possible examples come to mind. Transposable elements [30], integration and excision of λ-bacteriophage [31] and sequence rearrangements or recombination in immunoglobulin heavy chains [32]. All of these involve site-specific recombination, a mechanism proposed for both *P. anserina* mitochondrial DNA [19] and the petite mutation in yeast [20]. All known transposable elements are characterized by direct or inverted repeats at both transposon termini which are transferred with the transposon sequence. Neither of these types of sequences are noted in α-DNA. In λ-bacteriophage, there is a 15-bp common core sequence on both λ-sequences and the host bacterial DNA. No such homologous sequences are observed on α-DNA. The best analogy which can be drawn is with immunoglobulin DNA. Here, flanking the inserted segment there is a conserved 7 bp palindrome as well as a 9 bp sequence which is not the same 5' and 3' but is in the same relative position [32]. *Early* et al. [32] proposed that these conserved sequence elements serve as anchors holding the two distal ends of the immunoglobulin genes together for excision. This same type of structure is present in the α-DNA bordering the junction sites of excision. Finally, we ask how can the mobility of mitochondrial gene regions bring about senescence? This is a very difficult question which for now may be answered only in general terms, since we must deal with two interacting genetic systems, the mitochondrial and the nuclear. Loss of mitochondrial function itself could readily lead to cell death. Excision and amplified replication of these mitochondrial senDNAs could compete with or suppress normal mitochondrial DNA replication. Whether these plasmid-like molecules are transposed within the mitochondrial genome is not known since it is, in general, destroyed [10]. But we have shown that integration is possible into the nuclear genome. It may be that integration into the nuclear genome prevents the spread of the senDNA plasmid throughout the cytoplasm. A parallel for this would be the repression of autonomous bacteriophage DNA replication upon insertion into the bacterial chromosome. In the case of the mex mutant, α-DNA sequences are absent from the mitochondrial DNA but are contained within the nuclear genome. Just as with the occurrence of the lysogenic state, this is a rare event and would not measurably affect the ageing of the entire culture. This implies that the function provided by the mitochondrial sequences absent from mex 'mutants' would be furnished by the nucleus. Such a state is not without precedence [33].

Even though provocative, the analogies we have drawn between cellular ageing, mobile elements and lysogenic-like repression requires further work. The determination of the nuclear DNA sequences flanking the inserted mitochondrial plasmid sequences will enable further comparison with other site-specific recombination mechanisms. It will be particularly interesting to ascertain whether these senDNAs code for a product necessary for transposition. With such studies, it is quite possible that we will have a clear understanding of cellular ageing in *P. anserina*. How these studies will apply to other senescent systems remains to be investigated.

References

1. Rizet, G.: Sur la longévité des souches de *Podospora anserina*. C. r. hebd. Séanc. Acad. Sci., Paris *237:* 838–840 (1953).
2. Rizet, G.: Les modifications qui conduisent à la sénescence chez *Podospora anserina* sont-elles de nature cytoplasmique? C. r. hebd. Séanc. Acad. Sci., Paris *244:* 663–665 (1957).
3. Marcou, D.: Notion de longévité et nature cytoplasmique du déterminant de la sénescence chez quelques champignons. Ann. Sci. Nat. Bota. *11:* 653–764 (1961).
4. Holliday, R.; Huschtscha, L.I.; Kirkwood, T.B.L.: Cellular aging. Further evidence for the commitment theory. Science *213:* 1505–1508 (1981).
5. Marcou, D.; Schecroun, J.: La sénescence chez *Podospora anserina* pourrait être due à des

particules cytoplasmiques infectantes. C. r. hebd. Séanc. Acad. Sci., Paris 248: 280–283 (1959).

6 Smith, J.R.; Rubenstein, I.: Cytoplasmic inheritance of the timing of senescence in *Podospora anserina*. J. gen. Microbiol. 76: 297–304 (1973).

7 Tudzynski, P.; Esser, K.: Inhibitors of mitochondrial function prevent senescence in the ascomycete *Podospora anserina*. Mol. gen. Genet. 153: 111–113 (1977).

8 Belcour, L.; Begel, O.: Mitochondrial genes in *Podospora anserina*: recombination and linkage. Mol. gen. Genet. 153: 11–21 (1977).

9 Cummings, D.J.; Belcour, L.; Grandchamp, C.: Mitochondrial DNA from *Podospora anserina*. I. Isolation and characterization. Mol. gen. Genet. 171: 229–238 (1979).

10 Cummings, D.J.; Belcour, L.; Grandchamp, C.: Mitochondrial DNA from *Podospora anserina*. II. Properties of mutant DNA and multimeric circular DNA from senescent cultures. Mol. gen. Genet. 171: 239–249 (1979).

11 Cummings, D.J.; Laping, J.L.; Nolan, P.: Cloning of senescent mitochondrial DNA from *Podospora anserina*: a beginning; in Kroon, Saccone, The organization and expression of the mitochondrial genome, Bari 1980, pp. 97–102 (Elsevier/North Holland/Biomedical Press, New York 1980).

12 Jamet-Vierny, C.; Begel, O.; Belcour, L.: Senescence and *Podospora anserina*: amplification of a mitochondrial DNA sequence. Cell 21: 189–194 (1980).

13 Kuck, U.; Stahl, U.; Esser, K.: Plasmid-like DNA is part of mitochondrial DNA in *Podospora anserina*. Curr. Genet. 3: 151–156 (1981).

14 Mannela, C.A.; Goewert, R.R.; Lambowitz, A.M.: Characterization of variant *Neurospora crassa* mitochondrial DNAs which contain tandem reiterations. Cell 18: 1197–1209 (1979).

15 Lazarus, C.M.; Earl, A.J.; Turner, G.; Kuntzel, H.: Amplification of a mitochondrial DNA sequence in the cytoplasmically inherited 'ragged' mutant of *Aspergillus amstelodomi*. Eur. J. Biochem. 106: 633–641 (1980).

16 Wright, R.M.; Laping, J.L.; Horrum, M.A.; Cummings, D.J.: Mitochondrial DNA from *Podospora anserina*. III. Cloning, physical mapping, and localization of the ribosomal RNA genes. Mol. gen. Genet. 185: 56–64 (1982).

17 Southern, E.M.: Detection of sequences among DNA fragments separated by gel electrophoresis. J. molec. Biol. 98: 503–517 (1975).

18 Maxam, A.; Gilbert, W.: A new method for sequencing DNA. Proc. natn. Acad. Sci. USA 74: 560–564 (1977).

19 Belcour, L.; Begel, O.: Lethal mitochondrial genotypes in *Podospora anserina*: a model for senescence. Mol. gen. Genet. 163: 113–123 (1978).

20 Bernardi, G.: The petite mutation in yeast. Trends Biochem. Sci. 4: 197–201 (1979).

21 Wright, R.M.; Horrum, M.A.; Cummings, D.J.: Are mitochondrial structural genes selectively amplified during senescence in *Podospora anserina*? Cell 29: 505–515 (1982).

22 Macino, G.: Mapping of mitochondrial structural genes in *Neurospora crassa*. J. biol. Chem. 255: 10563–10565 (1980).

23 Belcour, L.; Begel, O.; Mosse, M.-O.; Vierny, C.: Mitochondrial DNA amplification in senescent cultures of *Podospora anserina*: variability between the retained, amplified sequences. Curr. Genet. 3: 13–22 (1981).

24 Lazdins, I.; Cummings, D.J.: Autonomously replicating sequences in young and senescent mitochondrial DNA from *Podospora anserina*. Curr. Gen. 6: 173–178 (1982).

25 Wright, R.M.; Cummings, D.J.: Integration of mitochondrial gene sequences coding for

subunits I and III of cytochrome c oxidase within the nuclear genome during senescence in *Podospora anserina.* Nature, Lond. *301:* 86–88 (1983).

26 Vierny, C.; Keller, A.M.; Begel, O.; Belcour, L.: A sequence of mitochondrial DNA is associated with the onset of senescence in a fungus. Nature, Lond. *297:* 157–159 (1982).

27 Cummings, D.J.; Wright, R.M.: DNA sequence of the excision sites of a mitochondrial plasmid from senescent *Podospora anserina.* Nucl. Acids Res. *11:* 2111–2119 (1983).

28 Pall, M.L.: Gene-amplification model of carcinogenesis. Proc. natn. Acad. Sci. USA *78:* 2465–2468 (1981).

29 Yadav, N.S.; Vanderleyden, J.; Bennett, D.R.; Barnes, W.M.; Chilton, M.-D.: Short direct repeats flank the T-DNA plasmid on a nopaline Ti plasmid. Proc. natn. Acad. Sci. USA *79:* 6322–6326 (1982).

30 Calos, M.P.; Miller, J.H.: Transposable elements. Cell *20:* 579–595 (1980).

31 Landy, A.; Ross, W.: Viral integration and excision: structure of the lambda att sites. Science *197:* 1147–1160 (1977).

32 Early, P.; Huang, H.; Davis, M.; Calame, K.; Hood, L.: An immunoglobulin heavy chain variable region gene is generated from three segments of DNA: V_H, D and J_H. Cell *19:* 981–992 (1980).

33 Tzagoloff, A.: Mitochondria (Plenum Press, New York 1982).

D.J. Cummings, PhD, Department of Microbiology and Immunology,
University of Colorado Health Sciences Center, Denver, CO 80262 (USA)

Subject Index

Acetylcholine
  of Alzheimer brain 50, 51
  of normally aging brain 44, 47, 48
Acetylcholine esterase 49
Actin 246
Age pigments 69, 210, 217
  of mouse neuroblastoma cells 210, 217
  of retinal pigment epithelial cells 81, 83, 88, 90, 91
Aging, cellular
  adaptive theory 10, 11
  amino acid transport 234–242
  autofluorescence as index 227
  band 3 degradation 245–253
  biology 10–12
  commitment theory 16, 17, 70–72
  damage theories 14, 15
  definition 10, 21, 63
  within different organs and tissues 13
  difficulties in study 2
  disposable soma theory 12, 15
  DNA polymerase 69
  environmental factors 122, 126
  error theory 65–67, 72, 73, 221, 231
  evolutionary perspective 10–12, 42
  genetic control 5, 15, 64, 65, 159
  Hayflick's experimental model 61, 62
  historical perspective on study 9
  immunological decline and 122
  in vitro 3, 4, 80–106, 193–195
  in vivo 3, 4, 80–106
  limited replication reflective of 13
  mechanisms 4–6, 14–16, 158–176
  membrane changes 234
  mutation theory, somatic 15, 64, 65
  vs. organismal aging 3, 9, 14, 18
  and population dynamics 193–205
    cell hybridization studies 196–199
    of human diploid fibroblast cultures 194–196
    immortal × SV40-transformed cell fusions 201, 202
    mechanism 202–205
    normal × immortal cell fusions 200, 201
    normal × normal cell fusions 199, 200
  premature
    paromomycin-induced 221–232
    in Werner's syndrome 64
  programme theories 5, 14, 15, 63, 64
  of retinal epithelium, human 80–91
    cell cultures for study 81, 82, 85–89
    electron microscopic study 82–85
    materials and methods for study 81, 82
    protein patterns 82, 89, 90
  in semelparous vs. iteroparous species 10
  single-celled organisms as models 6
  species specificity 21
  stochastic nature 5, 14, 63
  unbalanced growth 62
  unified theory 70–73
Aging, organismal 12–14

Subject Index

(Aging, organismal)
  brain, see Brain, aging
  cellular aging linked to 3, 9, 14, 18, 94, 178
Aldolase 50, 245
Alzheimer's disease 42
  biochemistry 46, 47, 49–54
    cellular and subcellular changes 49, 50
    cholinergic neurons and 50–54
    receptor cells 51, 55, 56
  therapeutic intervention 55
Amino acids
  activity
    in aging brain, normal 45, 47, 48
    in Alzheimer brain 50–52
  transport, with aging 234-242
γ-Aminobutyrate
  of Alzheimer brain 51
  in normally aging brain 45, 46
Animal models for aging experiments
  chick 60, 61
  mice 23–26, 210–218
  rat 43, 47, 48, 167–174
  single-celled organisms 6
Antibodies
  age-associated defects in responses to 116, 117
  anti-idiotype 116, 118, 119, 127, 132, 133
  to band 3 degradation products 249
  high-affinity vs. low-affinity 115
  interleukins affecting 125
  production, with age 112, 114, 115
  to terminal transferase 152
  for thymus characteristics identification 144
Antigens 115
  in autoimmunity 116
  blood group 122
  for lymphocyte differentiation in thymus 146, 149–151
  of major histocompatibility complex 142–149
  presentation, in Peyer's patches 129, 130
  senescent cell 245
Antioxidants 217
Aspergillus nidulans 255
Autoantibodies 112, 116

Auto-anti-idiotypic antibodies 132, 133
Autoimmune disease 110, 111
Autoimmunity 116, 125
Autoradiography
  of band 3 degradation products 249
  of DNA repair 97
Azobenzenearsonate 30

B cell growth factor 125
B lymphocytes
  aging T lymphocytes' effect 30
  in antibody production, with age 114
  autoimmune 113
  competitive repopulation assay 29, 31
  defective functioning, with age 116, 117, 123
  deficiencies 126
  differentiation 111
  mitogen-induced proliferation 125
  peripheral populations 111, 112
  of Peyer's patches 129, 130
  proliferative capacity, with age 123
  receptor/idiotype repertoire, with age 126–128
Band 3
  degradation, with aging 245–253
  extracellular space communication 249
  senescent cell antigens from 246
Bicarbonate 245
Biology of aging 10–12, 21
Blood group antigens 122
Brain, aging
  biochemistry
    normal 43–47
      amino acid transmitter synapses 45–47
      cholinergic interneurones 47, 48
      preservation of synaptic structures and 48
      receptor cells 45, 47, 48
      weight loss and atrophy 42, 43
    with Alzheimer's disease 49–54
      cellular and sub-cellular changes 49, 50
      cholinergic neurons, with long axons 50, 51

Subject Index

receptor cells 51, 55
cholinergic neurones, selective
vulnerability 42–56
therapeutic intervention 55, 56
Bruch's membrane 80, 83–85
Burst-forming units 29

Cancer
from chromosomal defects 5
extrachromosomally replicating
plasmid-like molecules 263
Cataracta senilis 96
Caudate nucleus
adrenal tissue transplantation 55
of Alzheimer brain 49
of normally aging brain 43–48
Cell
accuracy of germ vs. somatic 72, 73
cytoplasm, in aging 5, 6, 66
amino acid transport through 234–242
DNA synthesis inhibitor 196, 198
death, vs. cell population death 62, 63
division, vs. rate of organelle
formation 62
error frequencies, measurement 67–69
growth potential of heteroploid
transformed vs. diploid 60, 61, 64
incubation period 71
lens epithelial, aging 94–106
growth behavior 96, 97, 105
protein synthesis 99–103, 105
repair capacity 97–99, 105
mitochondria, in aging 62
in mouse neuroblastoma cell
cultures 210–218
in *Podospora anserina* 254–264
neoplastic
nuclear DNA-protein complexes 179, 180–185
propagation 178
nucleus, in aging 5, 66
chromatin, *see* Chromatin
mitochondrial DNA interaction
with 159–161
prokaryotic vs. eukaryotic 178
protein synthesis, with aging 15

replication 13, 60
retinal pigment epithelial 80
age-related pigment granules 80, 81
cultures 81, 82, 85–89
function 80
gel electrophoresis of protein
components 82, 89, 90
histology 80
light and electron microscopy 82–85
protein patterns 82, 89, 90
replication, after fetal development 80
in steady state or metastable condition 66, 72
Cell populations
of central nervous system, with age 42, 50
chick, cultured 60, 61
death, vs. cell death 62, 63
doublings, vs. life span 178
fibroblasts, *see* Fibroblasts
hemopoietic, *see* Hemopoietic stem cell
aging
highly vs. less differentiated 4
immortal 16, 17, 63
immune system 3, 28–30
MRC-5 and WI-38 63, 65, 69–71, 221–231
neoplastically transformed 178–180
neuroblastoma 210–218
population dynamics 193–205
replicative ability, vs. aging 13
size, vs. longevity 72
somatic hybrids 193, 196–202
via immortal × SV40-transformed cell
fusion 201, 202
via normal × immortal cell fusions 200, 201
via normal × normal cell fusions 199, 200
Cell-mediated immunity 112, 113
Central nervous system
increased life span, with evolution 42
neurons, cholinergic 3, 4, 43, 50–54
Chimaeric animals 14
Chloride 245
Choline 55
Choline acetyltransferase
in aging brain 44, 47, 48

Subject Index

(Choline acetyltransferase)
  in Alzheimer brain 49–53
Cholinergic interneurones 47, 48
Cholinergic neurones
  of Alzheimer brain 50–54
  of normally aging brain 3, 4, 43
Choriocapillaris 81, 85
Chromatin
  acetylation 168–170
  chemical composition 162
  compositional properties, with
    age 160–167
  condensation, in aging 187–188
  fractionation, for study 160, 161, 167
  genetic restriction, in aging 171, 172
  RNA initiation sites, for transcription 170, 171
  structure
    during meiosis 186, 187
    in normal aging vs. neoplastic
      cells 179–186
  sulfide-sulfide bonding, with age 180
  transcription 160, 167–174
  transcriptionally active vs. repressed
    portions 161, 162
Chromosomes
  markers, in repopulation assay studies 29, 31
  mutation, with aging 65
    of lens epithelial cells 97
    paromomycin suppression 221
  proteins 161–167
Colony-forming units 29
Commitment theory of aging 16, 17, 70–72
Computer simulation of cellular aging
  theory 203–205
Concanavalin A 125
Corpus striatum of aging brain 47, 48, 54
Crown gall disease 263
Crystallins 95, 96, 99–105
Cycloheximide 197
Cytolysosomes 217
Cytoplasm of cell
  amino acid transport through, with
    aging 234–242
  inhibitor for DNA synthesis, with
    aging 196, 198
  as source of aging 5, 6, 66

Dementia 42
  Alzheimer's disease and 50
  cellular and subcellular changes in
    brain 49
  therapeutic intervention 55
Deoxyribonuclease 160
Dinitrophenylated bovine gamma
  globulin 115
Disposable soma theory of aging 12, 15
DNA in aging
  chromosomal proteins affecting
    expression 165
  damage, with aging 187
  errors 15, 62, 66
  of eukaryotic vs. prokaryotic cells 179
  fractionation, for study 160
  mitochondrial 62, 254–264
    homology 256–259
    materials and methods of study 255
    nuclear interaction with 259–261
    restriction enzyme mapping 257, 258
    results of study 255–262
    sequence, at excision sites 262
  repair
    capacity 64
    in lens epithelial cells 97–99, 105
    meiosis and 178, 179
  synthesis, inhibition by senescent
    cells 196, 198
  transcription 159, 161
DNA polymerase 69, 152
DNase I 180
DNase II 160, 161, 167
Dopamine decarboxylase of aging brain 48
Drusenoid inclusions 81

Electron microscopy
  of aging pigments 210
  of aging retinal pigment
    epithelium 82–85
  of DNA complexes of normal aging vs.
    neoplastic cells 186
  of mitochondrial alterations in senescent
    cells 211
Electrophoresis, gel
  of band 3 degradation products 248, 249
  of chromosomal proteins 163, 164

Subject Index

of DNA-protein complexes of normal vs.
neoplastic cells 181
of inhibitor protein against DNA
synthesis 197
of lens epithelial cell proteins 99
of nuclear vs. mitochondrial DNA with
aging 260, 261
of protein patterns of aging retinal
pigment epithelium 82
Epithelial cell aging
lens 94–106
retinal pigment, human 80–91
thymic cortical and medullary 142, 145,
147
Error theory of cellular aging 65–67, 72, 73,
221, 231
Erythrocytes 29, 122, 245
Erythropoietic cells
competitive repopulation assay 29, 31
hemoglobin markers for determination of
function 33, 34
transplantation effects 32, 33
Erythropoietin 29
Escherichia coli 66, 67
Eukaryotic cells 178
Evolutionary perspective of aging 10–12
central nervous system 42
error theory 72

Fibroblasts
amino acid transport, with aging 234–242
of Bruch's membrane 85
clonal variability 194, 195
differentiation of cultured 61, 62
heterogeneity in growth potential among
individual 63
MRC-5 63, 65, 69–71, 221–231
paromomycin-induced aging 221–232
population dynamics 193–196
proliferative capacity, finite 13, 14, 60–62,
94
protein error frequencies 67, 68
stochastic features of aging 14, 72
tetraploid vs. diploid, longevity 65
virus infection, error frequency 68
5-Fluorouracil 69
Free radical-induced reactions 217

Fungal mitochondrial DNA with
aging 254–264

Gammopathies, monoclonal 112
Ganglioside content
of Alzheimer brain 49
of normally aging brain 43, 45
Genes
age-dependent modifications in
expression 159
chromosomal proteins in expression 165
in control of aging 5, 15, 64, 65
for immortalization, cellular 64
for inhibition of DNA synthesis 203, 204
mitochondrial, with aging 255–264
in oncogenesis 5
transcription 159, 160
translation 221
Germ cells
accuracy 72, 73
birth defects 17
immortal lineage 16, 17, 70
Glial cells in Alzheimer's disease 50
Glucose 6-phosphate 69
Glucose 6-phosphate dehydrogenase 222,
227, 228
Glutamate uptake, cellular 242
Glutathione 827
Glyceraldehyde-3-P-dehydrogenase 245
Glycophorin 246
Graft rejection 112, 113
Granulocytes 29

Haptens 115
Hassall's corpuscles 149
HeLa cell line 16, 196, 197
Hemoglobin
band 3 binding 245
markers, in competitive repopulation
assay studies 29, 31, 33–36, 38
Hemopoietic stem cell aging
background 21, 22
competitive repopulation assay in
study 30–33
with chromosome markers 29, 31
with hemoglobin markers 33–36
current research 37–39

Subject Index

(Hemopoietic stem cell aging)
  design of experiments 23–26
  immune system 28–30
  serial transplantations in
    experiments 26–28
Hexokinase 50
Histidinol 68
Histones 161
  acetylation 168
  electrophoretic separation 163, 164
  during meiosis 187
Hormones, thymic 111
Humoral immunity 112, 115
Huntington's chorea 43
Hybridization studies, cell 196–202
  immortal × SV40-transformed cell
    fusions 201, 202
  normal × immortal cell fusions 200, 201
  normal × normal cell fusions 199, 200
Hydrogen peroxide 218
Hypersensitivity reactions 112

IgA 112, 129
IgG 112, 115, 117, 129, 245, 247
IgM 112, 129
Immortal cells 16, 17
Immune system, aging
  B lymphocyte, *see* B lymphocytes
  cell-mediated immunity 112, 113
  cellular basis 113–119
  competency 3
  endocrine function of thymus 111
  humoral immunity 112
  idiotype repertoire changes 122, 126–128
  in lymph nodes 131, 132
  lymphocyte differentiation and 110, 111
  at mucosal lymphoid system level 123, 128–132
  in mucosal-associated lymphoid system 129
  in Peyer's patches 129, 130
  reduced heterogeneity 132, 134
  stem cells 28–30
  T lymphocyte, *see* T lymphocytes
  thymus involution and 110
Immunoelectrophoresis
  of band 3 degradation products 248, 249

  of lens epithelial cell proteins 99–104
Immunoglobulins 112
Incubation period for cell 71
Intellect 42
Interleukins 114, 124, 125
Interneurones, cholinergic 47, 48
Iteroparous species 10–12

Lecithin 55
Lens 95
  nucleus and cortex 95
  opacities 96
  proteins 95, 96
Lens epithelial cells, aging 94–106
  growth behavior 96, 97, 105
  protein synthesis 99–103, 105
  repair systems 97–99, 105
Leucine uptake, cellular 241
Life history organization 11
Life span
  cell damage repair vs. 15, 72
  cellular aging as determinant 13
  of fetal lung strains, MRC-5 and WI-38 63
  of germ line lineages 70
  hemopoietic stem cell 27
  of lens epithelial cells 96
  paromomycin effect 222–226
  population doubling and 178
  population size affecting 72
  of somatic cell hybrids 198
  of somatic tissues 70
  species vs. in vitro 193
Light microscopy of aging retinal
  epithelium 82
Lipofuscin
  of mouse neuroblastoma cells 210, 217
  of retinal pigment epithelial cells 81, 83, 88, 90, 91
Lymph nodes 131, 132
Lymphocytes
  B, *see* B lymphocytes
  differentiation 110, 111
    antigens 149–151
    materials and methods for study 143
    results of study 143–151
    terminal transferase 151

Subject Index

in thymus 142–153
leukemic 184
proliferation, with aging 188
senescent cell antigens 245
T, see T lymphocytes
Lymphoid system, mucosal-associated 122, 128–130
  immune dysfunction 131
  immune response 129
  microenvironment 129
Lymphokines 125

Macrophages
  in immune response 115, 125
  for senescent cell antigen removal 245
Major histocompatibility complex antigens 142–149
Malonaldehyde 218
Meiosis
  chromatin structure during 186, 187
  rejuvenation associated with 178
  repair of DNA and 178, 179
Melanin 80, 81, 83
Microperoxisomes 217
Mitochondria in cellular aging 62
  DNA 254–264
    at alpha excision sites 262
    homology of senescent 256–259
    materials and methods for study 255
    nuclear interaction with 258–261
    restriction enzyme mapping 256–259
    results of study 255–262
  of neuroblastoma cells, mouse 210–218
    decreased number 213
    energy-dependent processes 217
    free-radical peroxidation 213, 217
    materials and methods for study 211
    results of study 211–213
    structural alterations 213
    superoxide 218
Mitogens 113, 114, 125
Mutations in cellular aging 15, 64, 65
  of lens epithelial cells 97
  paromomycin suppression 221

Natural selection 11
Neocortex
  of Alzheimer brain 46, 49, 51, 54
  of normally aging brain 43–45, 48
Neoplastic cells
  nuclear DNA-protein complexes 179, 180, 181
  propagation 178
Neuroblastoma cells, aging
  lipofuscin accumulations 210, 217
  mitochondrial alterations 210–218
Neuronal aging
  cholinergic
    with Alzheimer's disease 50–54
    normal 3, 4, 43
  compositional properties of chromatin 161–167
  genetic control 158–176
  of neuroblastoma cells 210–218
  transcription of chromatin i 167–174
Neurospora crassa 66, 259
Nuclear matrix or scaffold 179
Nucleases 180
Nucleoside transport, cellular 241
Nucleosomes 160
Nucleus of cell
  chromation, see Chromatin
  mitochondrial DNA interaction with, in aging 258–261
  as source of aging 5

Oncogenes 5
Oncogenesis 5
Organismal aging
  of brain, see Brain, aging
  cellular aging linked to 3, 9, 14, 18, 94, 178
Osmiophilic bodies 211
Ouabain 199

Parkinsonism
  caudate nucleus 43
  therapeutic intervention 55
Paromomycin 69
  autofluorescence 227
  cellular morphology affected by 226–227
  glucose 6-phosphate dehydrogenase and 227–229
  growth and longevity affected by 222–226

Subject Index

(Paromomycin)
  premature aging 221–232
Peyer's patches 129, 130
Phosphofructokinase 245
Phosphogluconate dehydrogenases 6–69
Phosphohexoisomerase 50
Photoreceptor cells 80
Phytohemagglutinin
  interleukin response to 125
  receptors, on T lymphocytes 113
  for T cell stimulation 29, 31
Pigment granules
  of mouse neuroblastoma cells 210, 211, 217
  of retinal epithelium 80, 81
Plaque formation in central nervous system 42, 49
Plasma cells with age 123, 129, 130
Plasmids, DNA 263, 264
Platelets 29
Podospora 62, 72, 254–264
Polymerases
  DNA 69, 152
  RNA 159, 169, 170
Population dynamics 193–205
Progeroid syndromes, segmental 13
Prokaryotic cells 178
Proline uptake, cellular 240, 242
Protein
  band 3 245–251
  chromosomal 161
  content, in aging brain 45
  crystallins 95, 96
  in DNA complexes of normal aging vs. neoplastic cells 181
  errors, with aging 66, 67
  glucose transport 245
  histones 161, 163–165
  inhibitor, for DNA synthesis 197
  lens 95, 96
  in lens epithelial cell aging 94, 99–103, 105
  of retinal pigment epithelial cells 82, 89, 90
  of SV40-transformed cells 188
  synthesis, with aging 15, 65
  temperature effect 66

  transport, with aging 234
Prothymocytes 142
Protistan systems 6

Quinolinic acid 56

Radiation
  competitive repopulating ability affected 35, 36, 39
  of DNA of lens epithelial cells 98, 105
Respiration in mitochondria 217, 218
Retinal epithelium aging 80–91
  cell cultures in study 81, 82, 85–89
  electron microscopy studies 82–85
  gel electrophoresis in study 82, 89, 90
  light microscopy in study 82
  materials and methods in study 81–83
  protein patterns 89
  results of study 83–90
RNA
  DNA transcription 159
    with aging 167, 168
    elongation velocity 173
    neuronal content, with age 173
RNA polymerases 159, 169, 170

Self-tolerance in immunity 116
Semelparous species 10, 12
Senescence, *see* Aging
Senility 3
Serine uptake with aging 239, 242
Serotonin
  in Alzheimer brain 50
  in normally aging brain 44, 47
Somatic cells
  accuracy 72, 73
  hybrid 196–202
  immortalization 16
Somatic mutation theory of aging 15
Species
  lens epithelial cells of differing, life span 96
  semelparous vs. iteroparous 10–12
  specificity, for aging 21
Spectrin 245, 246
Stem cells
  differentiation 29

fetal vs. old and young 38, 39
hematopoietic, see Hemopoietic stem cell
   aging
of immune system 28–30
immunologic defects of age and 29
intrinsic timing of aging 21, 37
life span 3, 4
lymphohematopoietic activity, with
   age 123
Streptomycin 67
Succinate dehydrogenase 50
Sugar transport with aging 241
Superoxide dismutase 64, 218
SV40-transformed cells 188
   in cell hybridization studies of population
      dynamics 196, 198, 200–202
   DNA synthesis inhibitor negated by 196,
      197

T lymphocyte replacing factor 115
T lymphocytes
   in antibody production, with age 114
   autoimmune 113
   in cell-mediated immunity 113
   competitive repopulation assay 29, 31
   cytotoxic 113, 124
   differentiation 110, 111
   differentiation antigens 149–151
   helper 30, 116, 123, 124
   hypersensitivity reactions 123
   mitogen-induced proliferation 113, 114,
      125
   mixed leukocyte reactions 123
   mutation, frequency 65
   peripheral populations 111, 112
   of Peyer's patches 129
   phytohemagglutinin receptors 113
   phytohemagglutinin stimulation 31
   proliferative defect, with age 114
   receptor/idiotype repertoire, with age 126
   responsiveness, with age 30, 31, 115, 124,
      125
   suppressor 30, 117, 123, 124
Temperature effect on protein error 66
Temporal lobe
   of Alzheimer brain 46, 47, 49, 50, 54
   of normally aging brain 43–45, 48

Teratocarcinoma cells, malignant 17
Terminal transferase 151
Thioguanine 199
Thymectomy 118
Thymic hormones 111
Thymic nurse cells 142
Thymocytes 111
Thymopoietin 111
Thymosin alpha-1 111
Thymus
   changes, with aging 122
   endocrine function 111
   interdigitating reticulum cells 142, 148,
      149
   involution 110
   lymphocyte differentiation and
      microenvironment 142–153
      antigens 149–151
      materials and methods for study 143
      results of study 143–151
      terminal transferase 151
Tissue transplantation
   adrenal, to caudate nucleus 55
   in aging experiments 21, 22
   in Alzheimer disease treatment 55, 56
Transcription 159, 160
   adaptability, in aging 172
   aging effects 167–174
   chromosomal proteins 161, 165
   regulatory mechanisms 160
Transplantation experiments
   competitive repopulating ability affected
      by 32
   design 23–26
   for intrinsic vs. extrinsic timing
      determination of aging 21, 37
   pitfalls 21, 22
   serial 26–28
Tumor cells, see Neoplastic cells
Tyrosine hydroxylase in aging brain 47,
   48

Unitary immunological theory of aging 122

Viruses
   fibroblast error frequencies with infection
      by 68

(Viruses)
  SV40
    in cell hybridization studies of population dynamics 196, 198, 200–202
    DNA synthesis inhibitor negated by 196, 197
Vitamin A 80

Werner's syndrome 13, 64